MOVING FORWARD

GIS FOR TRANSPORTATION

Edited by
Terry Bills
Keith Mann

Esri Press
REDLANDS | CALIFORNIA

Esri Press, 380 New York Street, Redlands, California 92373-8100

Printed in the United States of America
25 24 23 22 21   1 2 3 4 5 6 7 8 9 10

Library of Congress Control Number: 2021943347

For purchasing and distribution options (both domestic and international), please visit esripress.esri.com.

On the cover: Photograph by Kalafoto.

CONTENTS

INTRODUCTION

SIMPLY STATED, TRANSPORTATION IS ABOUT THE MOVEMENT of people and goods from one place to another. However, history has shown that advancements in transportation have brought prosperity to emerging nations. Over time, new transportation technologies increased the speed and capacity of transport, dramatically altering economies. Distribution became more extensive and pervasive. Populations shifted geographically, drawn to new transport infrastructure. Even today, one might say that the importance of transportation continues to be a driving force behind how people live their lives and why businesses thrive or fail.

The ever-evolving development of transportation systems around the world carries with it the legacy of older systems. Transportation agencies struggle to maintain and modernize the remnant pieces of roads, subways, and deep-water ports. Preservation of iconic bridges, train stations, and other feats of engineering and architectural genius and beauty from bygone days is also a priority. And yet, the world is changing. Populations are skyrocketing in some places while dissolving in others. Demand is greater for next-day delivery. The promise of self-driving vehicles exists as does the threat of rising sea waters. The cost of fixing what is broken mounts. So many challenges must be faced.

In this book, the word *transportation* refers to the work that organizations and agencies—such as departments of transportation, airports, and transit and port authorities—do to plan, manage, and maintain the infrastructure that keeps people and things moving efficiently, safely, reliably every day, and into the foreseeable future. The term *transportation system* refers to all the pieces and parts, large and small, that make transport possible. It is a system that includes the people who monitor, repair, and build components of the system. Managing transportation systems is also about good business and customer relationships; preparing for and responding to incidents, such as an accident or weather-related events; and making sure that facilities and services are meeting the needs of communities, supporting supply chains, and driving local and global economies.

Nearly every aspect of a transportation system is locational, meaning that every element of the system has a place on the earth and can be mapped and analyzed spatially. Transportation agencies use geographic information system (GIS) technology to help improve operational efficiency, safety and security, asset management, and planning and sustainability. GIS provides a stronger sense of location intelligence across organizations, so that everyone—including transportation managers, engineers, planners, and repair crews—can develop a better understanding of where people and things are in relation to everything else. GIS analysis reveals unseen issues, vulnerabilities, and movement patterns that help transportation professionals work better together. Location intelligence helps people understand why things happen where they do and inspires new ways to keep moving forward.

This book is organized in four parts.

Part 1: Operational efficiency

Operational efficiency depends on everyone working better together, being ready for problems, and understanding how the system can run at peak performance. Today, reaching the goals of an efficient system relies heavily on data. For transportation agencies, GIS provides a collaborative environment where sharing and using data is presented in a locationally intelligent view, allowing managers and staff to see and understand the bigger operational picture within a real-world context. In this section of the book, a collection of stories shows how transportation agencies are using GIS today to manage daily tasks, monitor real-time information flowing in from sensors, and investigate the patterns and trends of operational performance to save time and money.

Part 2: Safety and security

Moving people, goods, and services across the globe is fraught with risk. With risk come rules and regulations designed to keep people safe and protect operational infrastructure, which is why transportation agencies emphasize safety and security measures—to reduce risk within a complex environment of moving parts. GIS technology is especially well suited for strengthening safety and security measures, giving personnel real-time situational awareness and coordination of resources. In this section, a collection of stories shows how transportation agencies are using GIS to make transportation systems safer and more secure, helping them prepare for, respond to, and recover from minor incidents to major disasters.

Part 3: Asset management

Transportation agencies manage millions of assets—the pieces and parts of a transportation system—ranging from tiny sensors to buildings. And every asset has a location. With GIS, transportation agencies can create a comprehensive asset inventory that includes the precise locations of all assets. In this section, a collection of stories shows how transportation agencies are using GIS to help maintenance crews and asset inspectors capture detailed information that automatically updates asset management systems, as well as document work, prioritize work orders, retask crews based on their locations and proximity to other issues and assets, and move from reactive to predictive maintenance.

Part 4: Planning and sustainability

Building transportation infrastructure can cost billions of dollars, which means that planning new infrastructure, such as a bridge, or extending existing infrastructure, such as a new transit line, requires extensive planning. Transportation agencies are concerned with the long-term sustainability and resiliency of the infrastructure and always trying to anticipate growth in business and the changing needs of customers. With GIS, transportation agencies get a unique geographic perspective on understanding current conditions and existing stresses on transportation systems. In this section, a collection of stories shows how transportation agencies are using GIS to plan changes to the transportation system while better understanding the needs of customers, patterns of economic development, and meeting state and federal requirements.

HOW TO USE THIS BOOK

THIS BOOK IS DESIGNED TO HELP YOU ADD SPATIAL reasoning to decision processes and operational workflows. It is a guide for taking first steps with GIS and applying locational intelligence to common problems. Using the information from this book can help you create a more collaborative environment within your department and throughout your organization. You can use this book to identify where maps, spatial analysis, and GIS apps can be helpful in your work and then, as a next step, learn more about those resources. At the end of the book, the "Next Steps" section provides you with basic strategic advice for applying GIS to transportation.

Learn about additional GIS resources for transportation by visiting the web page for this book:

go.esri.com/mf-resources

PART 1

OPERATIONAL EFFICIENCY

MANY BUSINESSES SEEK TO REDUCE COSTS WHILE ensuring or growing revenue. For transportation agencies, such as government transportation departments, airports, transit authorities, and ports, operational efficiency is focused on moving people, goods, and services at maximum capacity without disruption. However, transportation systems are intricate and complex constructions built to safeguard system users and deliver durable and reliable service for many years. For example, an international airport may handle hundreds of thousands of flights, tens of millions of passengers, and millions of tons of cargo every year. An airport such as this strives to minimize risks to aircraft, people, and cargo, while maintaining schedules and providing a level of reliable service that their customers depend on.

The application of GIS to transportation is playing an increasingly larger operational role by giving managers and workers a locationally intelligent view of how the system is performing daily, over time, and in real time. Transportation managers also use GIS to see the bigger operational picture, study passenger and cargo movement, and respond to extenuating events, such as weather conditions. Transportation operations centers are using GIS maps and dashboards to monitor current activities and access information from existing business systems.

There are three ways that transportation agencies most commonly use GIS to improve operational efficiency.

(1) Achieving real-time situational awareness

Real-time GIS allows users to simultaneously integrate, analyze, and display streaming data from many sensors, devices, and social media feeds. Operations managers can filter feeds and define analytics to achieve real-time situational awareness of the locations of events, people, vehicles, vessels, and aircraft. Maps and databases are automatically updated to reflect status, and key personnel are simultaneously alerted when stated thresholds are exceeded. Operational data can be combined with transportation infrastructure data to discover trends, patterns, and outliers.

(2) Measuring operational performance by location

Transportation operations managers use GIS to convey historical and real-time information by presenting location-based analytics using interactive data visualizations. GIS dashboards help operations staff understand the status of events, projects, and unfolding situations in real time. GIS provides location-based tools for tabulating performance metrics, monitoring operational activities, and visualizing key performance indicators for defined geographies. Performance analysis results can be shared within transportation agencies and with partners, customers, and the public through web applications or a GIS hub community website.

(3) Improving mobile workforce collaboration

Operations managers and field staff use mobile GIS applications to coordinate work across a transportation agency's holdings. Mobile GIS can be integrated with work management systems to reduce or eliminate paper-based workflows and provide everyone with access to the authoritative data they need to make accurate and timely decisions. Mobile workers can carry their maps and assignments on smartphones or tablets to stay organized, report progress, call for assistance, reduce errors, and boost productivity. With GIS, operations managers have a clear view of project status, priorities, and next assignments.

In the following selection of case studies, you will learn how a state department of transportation uses GIS to create a live, user-friendly travel information map service for public and commercial drivers. In another story, you will see how an international airport uses GIS to save money while becoming more efficient and digitally transform its operations to incorporate hundreds of datasets. You will also read how a government agency uses GIS to manage and maintain efficient navigation on the Mississippi River by targeting dredging operations and being more prepared to respond to flooding that might overwhelm riverbanks and levees. Finally, you will learn how another airport uses GIS and location intelligence to give airport staff a better understanding of how passengers interact with the facility and monitor physical systems.

INFORMING ROADWAY TRAVELERS

Wyoming Department of Transportation

WITH 6,800 MILES OF HIGHWAY, WYOMING SERVES AS A major corridor for commercial truck traffic—particularly on Interstate 80, which runs through the southern part of the state. It is one of the busiest routes in the United States for moving freight coast to coast.

Truckers, tourists, and locals alike can now navigate Wyoming's often unpredictable highways and byways with the help of real-time information from the Wyoming Travel Information Map, built by the Wyoming Department of Transportation (WYDOT) using ArcGIS® Web AppBuilder. On an average day, the map gets about 170,000 visits—and that number can climb to 4 million when there's a big storm.

"Road condition information in Wyoming is critical to safety," said Vince Garcia, the GIS/ITS program manager for WYDOT. "Wyoming gets severe winters with a long duration. We have the worst blowing snow conditions in the country."

WYDOT created the Wyoming Travel Information Map largely to help drivers plan their travel before they set out on the road. "Without this map, people coming from other states may not be well prepared," Garcia added.

The interactive map displays the state's current road conditions, construction areas, and advisories. Other map layers include web cameras on certain parts of the highway, rest area locations, size and weight restrictions, weather stations, truck parking areas, and the locations of variable speed limit signs.

With so many elements in one map, WYDOT didn't want to build things from scratch. "We preferred to configure instead of code and customize or extend where we [needed] it," said Ben Saunders,

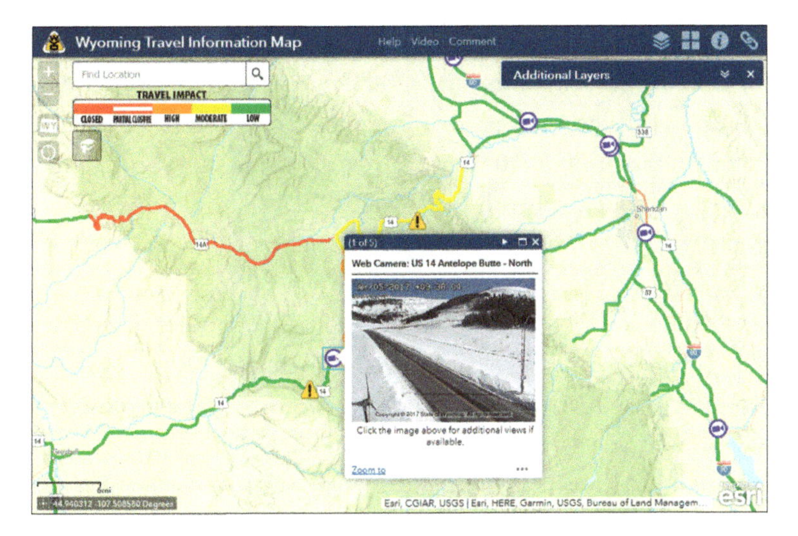

Users can turn on additional layers in the interactive map, such as web cameras on parts of Wyoming's highways.

a GIS professional with Srednaus Mapping, which worked with WYDOT to build the Wyoming Travel Information Map. "The functionality of ArcGIS Web AppBuilder is ready out of the box, [and] Esri's help documents and the user community gave us the confidence that we [were] working within a well-documented environment."

Web AppBuilder features ready-to-use widgets, configurable themes, and custom app templates. Apps can be hosted online or run on an on-premises server. And the HTML/JavaScript apps that it builds are responsive and intuitive.

"We wanted this map to be mobile-friendly, and [we] liked that ArcGIS Web AppBuilder offers a responsive design," said Trenton Rawlinson, a GIS developer with WYDOT. "We also wanted to make the initial download smaller so people can access it when there is limited internet service. We used image caching versus feature services to make it load more efficiently."

Although WYDOT created several iterations of the map during development, the basic process of building it took just three days. After tuning up its performance, administering user testing, and working through network issues, the WYDOT team delivered the Wyoming Travel Information Map to the public within six months of conceiving the project.

"There's no way we would have been able to put this together using the old technology," Garcia said. "Some of the lessons learned led directly to more efficient development of subsequent maps for the Wyoming Department of Tourism. And the platform allows us to add additional functionality and data, like National Weather Service radar loops. The application can continue to evolve."

This story originally appeared as "Navigating Wyoming's Snowy, Blowy Roads" in *ArcNews* (Winter 2018). Image courtesy of the Wyoming Department of Transportation.

EXPANDING AIRPORT CAPACITY

Geneva Airport, Switzerland

NOT LONG AGO, THE AIRPORT IN GENEVA, SWITZERLAND, faced an important crossroads: how to expand its operations to keep up with international travel. It had two major problems—tight space with little room for expansion and a tight budget. With only one runway and passenger numbers climbing, the airport was struggling to keep up with growth.

Further complicating the matter, Geneva is a center for tourism as well as international finance and home to many humanitarian groups, including the International Committee of the Red Cross and the European offices of the United Nations. The city is a gathering point for nations—a place where heads of state, diplomats, and dignitaries rush in and out and their planes often require priority scheduling and security that can disrupt normal airport flows.

Geneva Airport officials had only one option: improve efficiency by creating a centralized command post or nerve center. "The first practice that we worked on was to create a single operations room in which we have 30 workstations, which allow the various stakeholders across the airport to sit together," said Thomas Romig, head of the Airport Operations Center. The operations team included ground handlers, airside operations, landside operations, security personnel, and border control, as well as airlines and the airport's navigation service provider.

Team members needed to work together to find new efficiencies based on a common understanding of the data generated by the movements of planes, passengers, and baggage. They needed to create better processes for moving crowds to quicker access points and directing planes to different gates. Additionally, staff needed to quickly understand when and where passengers were waiting too

Inside the operations control room at Geneva Airport.

long in check-in or security lines, where those backups were likely to occur so they could be prevented, and where there might be a security breach or where a smoke alarm was activated, and how to respond. The data had to be readily understood by decision-makers, especially those without backgrounds in information technology.

Romig found support from Alexandre Pillonel, the GIS application manager in charge of the spatial data infrastructure in the airport's IT division. GIS technology excels at tracking the location of assets, from planes at an airport to delivery trucks on the road to assets across an electric grid.

Pillonel and his team built a GIS-based, real-time dashboard of airport operations from the point of view of the end users rather than the IT staff. Pillonel explained that a visualization of the daily operations through a map displays all the different locations of the airplanes. "We're integrating radar data into that so that we can see the movement of vehicles and aircraft on the same visual," said Pillonel about the improved safety and more efficient management of all traffic in the area.

For example, if a delayed flight keeps a plane parked at a gate that an incoming flight is scheduled to use, the dashboard registers an alarm. Immediately, all team members know about the conflict and can respond. The incoming flight is redirected and, if needed, airport and airline personnel are dispatched to the newly assigned gate, either by the authorized team members in the central situation room or via immediate communication with appropriate managers.

With more than 500 datasets being collected and analyzed, the dashboard receives information that includes security warnings, passenger flow, aircraft location, baggage movement, border and immigration controls, and the loading and unloading of airplanes. Dashboard watchers have "a common situational awareness of what's going on daily and pretty much in real time," Romig said.

The dashboard visualization extends to conditions on the roadways and railways leading to the airport, providing a sense of how passenger traffic ebbs and flows in the airport. As passengers enter the terminal and make their way to check-in areas and security lines, the internal system senses the locations of cell phones in operation. The system anonymizes that data as it tracks the locations and movements of groups of people. GIS is also used to track locations of planes on the ground and show managers if an aircraft is at a deicing station or being loaded or unloaded.

Thanks in part to the location intelligence provided by GIS, the airport currently handles 6 million more passengers a year than anticipated by the airport's original design. Romig notes that with 25 million passengers expected by 2030, the airport eventually will need to invest in major infrastructure upgrades to cater to future demand. However, he expects GIS to play an increasing role in helping to manage current and future growth efficiently.

In the meantime, the improved dashboard and GIS setup have eliminated significant costs, such as spending $60 million to build more taxiways or $300 million to build another terminal.

Romig believes that the Geneva approach can be applied in other industries as well. "I can imagine, in a big port or in some other manufacturing areas, where you've got parts all over a big surface—a big area—that GIS could certainly be useful," he said.

This story originally appeared as "Location Intelligence Saves Geneva Airport Millions" by Jeffrey Peters, in *WhereNext* (2017). Image courtesy of Geneva Airport.

KEEPING RIVER TRAFFIC FLOWING

US Army Corps of Engineers, Rock Island District

EACH FALL, BARGES LOADED WITH SOYBEANS AND CORN make their way from midwestern Corn Belt farms down the Mississippi River to the Port of New Orleans for export. As the world leader in grain production, the US relies on this river to carry 60 percent of its harvest, amounting to more than $600 billion in annual economic activity.

Any supply chain disruptions to barge traffic on inland waterways can have ripple effects on the country's economy. Mississippi River traffic must keep moving regardless of the pressing problems of aging infrastructure, drought, and flooding. Charged with managing the river is the US Army Corps of Engineers (the Corps), which provides flood control and maintains commercial navigation.

The Corps's Rock Island District operates 314 miles of channel and 12 lock and dam sites on the Mississippi, as well as 268 miles of channel and eight lock and dam sites on the Illinois Waterway.

Much of the Upper Mississippi infrastructure constructed in the 1930s has gone well beyond its engineered lifespan of 50 years. With more than $1 billion in deferred maintenance on Mississippi River locks and dams, the Corps must constantly monitor performance and make repairs to keep this infrastructure operational.

A recent study suggests that failure of any of the 25 aging locks on the Upper Mississippi River could result in nearly 500,000 truckloads to move 12 million tons of grain on highways between the Twin Cities and St. Louis. Trucking that load would cost hundreds of millions of dollars more than barge transport and would damage already stressed roadways.

In maintaining the Mississippi River, Corps engineers must also be vigilant of unpredictable weather.

Drought conditions in 2017, coupled with the biggest harvests in 29 years, triggered shipping backlogs and swelled grain storage facilities on riverbanks to capacity. This backlog led to record high freight costs for a mode that is reliably the cheapest way to move bulk commodities.

Then in March 2018, high water volumes from heavy rains disrupted navigation at multiple locations. This flooding did not compare to the record floods of 2008 or 1993, but it impacted barge traffic. The Corps reduces the length of barge tows during floods, often restricts traffic to daylight hours as opposed to usual 24/7 operations, and can, if needed, close the river to traffic.

The Rock Island District is at the crossroads of river transport in the country. It maintains crucial connections for goods going down the Mississippi to the Gulf of Mexico and up the Illinois Waterway to the Great Lakes and on to the Northeast by way of the St. Lawrence Seaway and the Erie Canal.

To constantly monitor and manage hundreds of miles of rivers, the Corps uses GIS to record, analyze, and visualize data about the rivers and the forces that affect their flow. They can then use GIS to share this information internally and with the public via interactive maps.

Inland waterway maintenance differs greatly from the Corps's work on fixed channels at coastal ports. River levels can vary by a depth of more than 30 feet over the course of a year. Yet the Corps is mandated to ensure these ever-shifting navigation channels remain as intended at 9 feet deep and 300 feet wide.

"You kind of have two rivers: the water and the sand on the river bottom, and they're both moving," said Dan McBride, geographer at the US Army Corps of Engineers, Rock Island District. "I would liken shoaling [the accumulation of sand] in a deep-draft port channel to a pothole that slows traffic and reduces cargo loads but

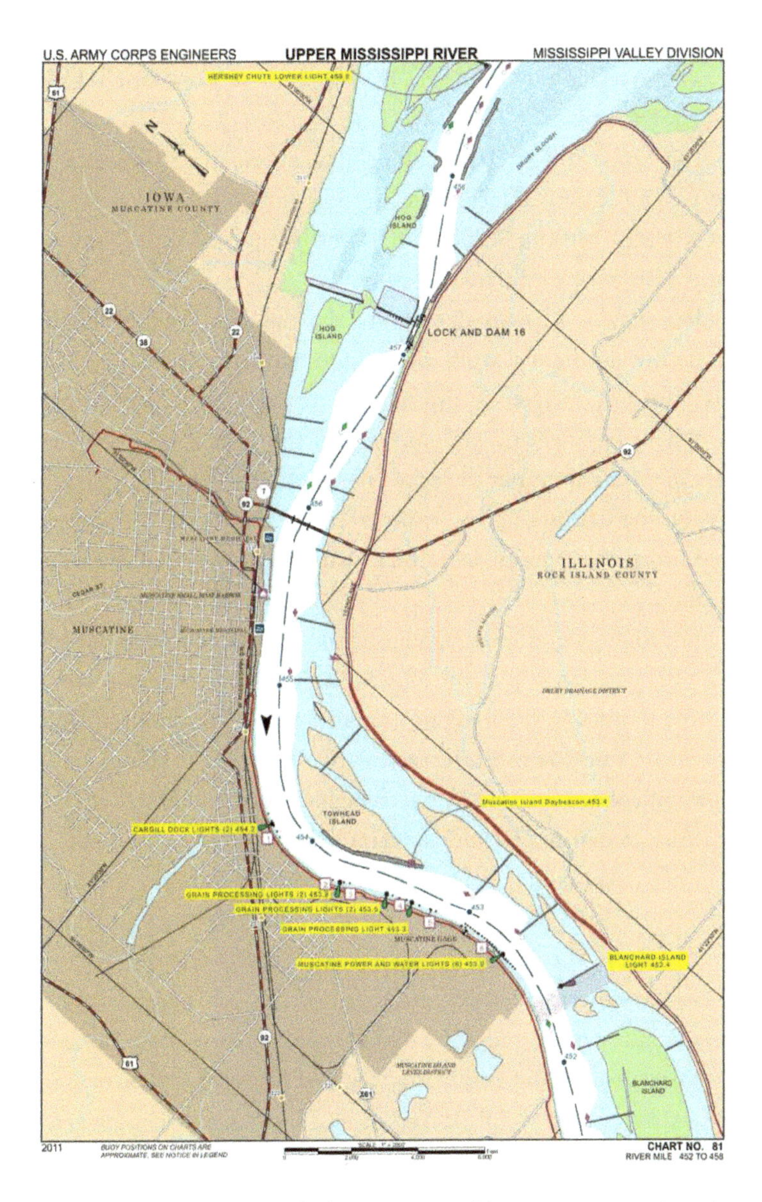

Inland Electronic Navigational Charts are viewable on monitors and can be downloaded as PDF documents. This navigational chart shows a section of the Mississippi River near Muscatine, Iowa.

shoaling in a shallow-draft river channel is more like a sinkhole that shuts down a road."

The Corps and its contractors conduct regular hydrographic surveys using a variety of measurement tools and sensors aboard special survey boats. Sensors include sonar sensors that use sound waves to reveal a 3D view of the river bottom.

The surveys identify changes to the channel, such as shoaling that constricts channel width or buildup that reduces channel depth. This information factors into immediate dredging action or more permanent engineering actions such as the construction of check dams in river tributaries to reduce their force. In addition to regular surveys, the Corps gathers details from US Geological Survey gauging stations that supply data about water flow and the force of any flooding.

"On the navigation side, flooding drops sediment at the confluence of rivers. It's important to get out there and survey as soon as possible to inform the dredging operations and restore the channel," said Tony Niles, assistant director for Civil Works Research and Development at US Army Corps of Engineers headquarters. "On the flood risk management side, the data need is different. The purpose isn't about maintaining depth, it's to see where flood waters are, how high they are, and the potential to overwhelm riverbanks and levees."

Recently, the Rock Island District started using an enterprise-wide system called eHydro that includes tools and workflows to catalog, organize, and share hydrographic surveys. By turning to a GIS-based system, and automating the process using Python scripting, the Corps can now rapidly turn data into a visual product for dredge coordinators to review.

With all its surveying work and a growing need to share details with vessel captains, the Corps was given authority by Congress in 2001 to maintain navigation charts for the country's inland waterways. Data the Corps regularly collects for waterway maintenance and flood control activities provides input to make the Inland Electronic Navigational Charts (IENCs).

Coast Guard regulations require all vessels to carry charts, and electronic charts have an advantage over paper.

"Electronic charts offer increased detail and the ability to offer rich data instead of just a symbol on paper," McBride said. "They also allow far more frequent updates, ensuring the charts on a vessel are up to date."

"We can make charts across all 15 inland districts for 8,000 miles of inland waterways that are all consistent, timely, and done as a single Corps-wide product as opposed to district by district," Niles said.

As the IENC program has progressed, data products have improved, and more captains have shifted to the digital system.

"Our charts were pretty rough at first, but they have had several major refreshes," McBride said. "Boat pilots now tell me that our charts are excellent and are viewed as gospel. It's humbling to have the responsibility to produce something that is trusted so much."

The Corps continues to look to sensors and systems to address the complexity of its challenges. It has recently tapped into signals from the automatic identification system (AIS) the Coast Guard maintains, receiving the signals that every vessel transmits to avoid collisions.

"The Corps has been tracking waterway usage for some time and has excellent statistics about how much cargo is transported on the nation's waterways," McBride said. "AIS data allows us to see vessel traffic at a granular scale. We've started comparing our recommended sailing line against the most traveled paths to reveal discrepancies. It's one more tool in our toolbox to inform maintenance decisions."

Corps researchers are looking at AIS data for other purposes. One team is looking to obtain real-time AIS signals to understand incoming traffic and improve the time it takes to get through lock

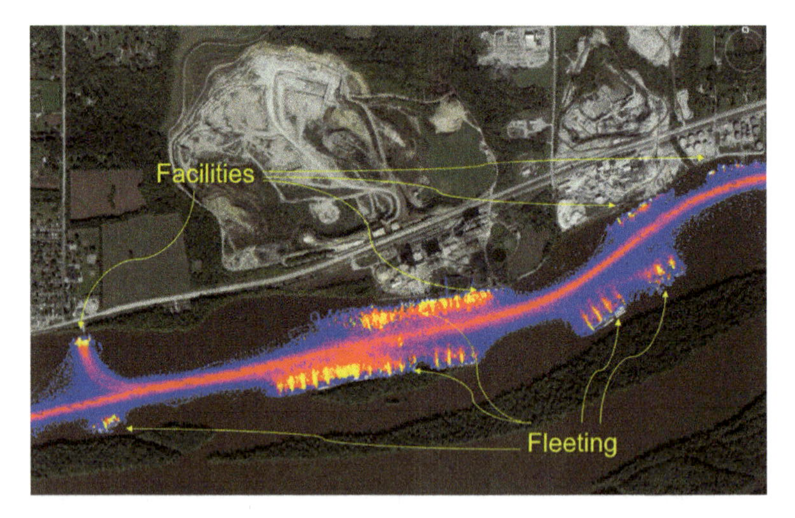

Visualizing historical AIS data reveals patterns. This heat map of a river terminal in Iowa shows the use pattern for loading aggregate and liquid fertilizer.

sites. Another team hopes to use AIS to both receive and send information to vessels. Another team is looking at the data to understand typical cargo trips and the time it takes for trip completion. The AIS is enabling the development of the River Information System (RIS) that will make real-time information available to vessels, waterway operators, and service providers for safer and more efficient navigation.

"With AIS data, we're seeing activity," McBride said. "It confirms things about the behavior of barge traffic and reveals more insight into local traffic versus those that are traversing through. My feelings are that we have just started to scratch the surface of what's possible."

This story originally appeared as "Sensors and Systems Keep Mississippi River Traffic Flowing" by Steve Snow, in the Esri Blog (2018). All images courtesy of the US Army Corps of Engineers and Esri.

THINKING BIG, WORKING FAST

Dublin Airport, Ireland

MORE THAN 31 MILLION PEOPLE PASS THROUGH DUBLIN Airport annually. The state-owned commercial company that oversees operation and management of Dublin Airport, daa, decided to digitally transform the facility so it could better serve customers and operate more efficiently. As part of this sweeping digital change, daa implemented the Esri ArcGIS system, which includes GIS dashboards, mobile data collection apps, and indoor mapping capabilities.

"Our aim was to think big, start small, and work fast," said Neil Moran, Dublin Airport's head of digital asset management and transformation. "ArcGIS gave us the platform we needed to connect systems, give all our employees a single view of the truth, and empower them to work more efficiently and flexibly all around the airport."

ArcGIS connects with daa's data systems, creating a single view of enterprise asset information, providing insight into the airport's operational costs, risks, and performance, and helping passengers move from parking lots to terminal gates more quickly and smoothly. Staff can track reports of incidents, such as an injury to a baggage handler. Crews can instantly capture information about airfield conditions, such as a problem with the pavement, so repairs can be made before the issue becomes hazardous.

Location intelligence gives airport staff a better understanding of how passengers interact with the facility and where issues need to be addressed. It tracks and monitors many of the facility's business assets such as lifts, escalators, parking lots, security, baggage systems, and boarding gates.

"We need to understand how passengers are interacting with us as they are traveling through the airport," said Vincent Harrison, Dublin Airport's managing director. "This information is increasingly important as we move more passengers through the airport. GIS shows us where issues need to be addressed and how we can respond to them very quickly."

To map the Wi-Fi signal patterns in Dublin Airport, daa works with Apple to deliver up-to-date information relevant to passengers using mapping apps on their devices. ArcGIS® Indoors™ is used in the Dublin Airport App, the mobile app that daa created for passengers. Passengers use the app as they would a car navigation system, which shows current location, provides a route to a boarding gate, and estimates the time it will take to get there. ArcGIS Indoors is also used on map kiosks within the airport, as well as desktops and mobile mapping apps for staff.

Dublin Airport's GIS also consumes data from the airport asset management system and design files, weather services, and other resources. The airport uses GIS to create operational visualizations of the data and applies analytics to generate airport intelligence. For

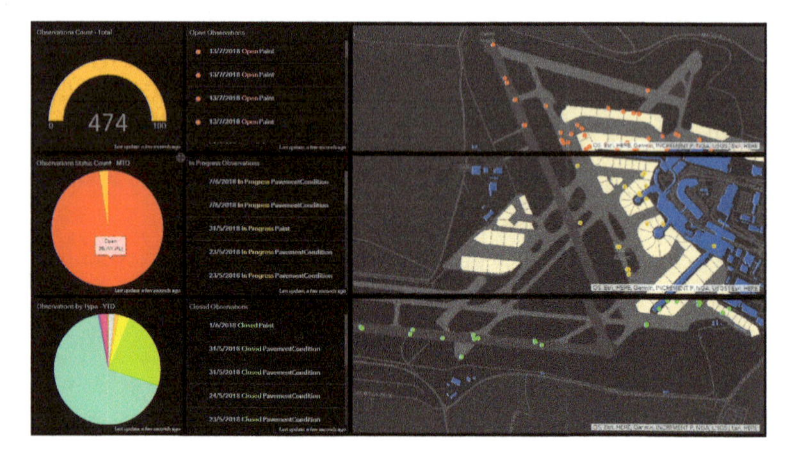

Dublin Airport tracks maintenance issues on runways using different types of GIS dashboards, such as this one, showing open and closed maintenance observations.

example, maintenance crews use ArcGIS to collect information about and monitor facility repairs. Airport workers use ArcGIS Collector and ArcGIS Survey123 to capture detailed information and photos about problems. Incidents appear on a browser-based dashboard using ArcGIS Dashboards. The dashboard processes the information and displays it in maps, graphs, and charts, giving managers immediate information about what is happening and which repairs to prioritize, schedule, and assign to technicians to complete.

The dashboard also tracks a task status. The maintenance technician receives an assignment, sees its location, and clicks the map to access the initial notes and photos. The technician can also use the app to access repair manuals. Upon completion, the technician marks the task complete in the app, and the task is instantly updated on the dashboard. The record of an asset's maintenance history is also accessible at any time and can be used for analyzing asset performance.

"Dashboards put real-time information into managers' hands and provide valuable insight into the airport's operations, risks, and performance, which we can use to make faster, well-informed decisions," said Morgan Crumlish, Dublin Airport's spatial data manager.

daa collects detailed incident information and uses it to speed response and drive health and safety improvements. For instance, if a passenger is injured on an escalator, staff members use ArcGIS Survey123 on their mobile devices to access and complete an accident report. The same app is used to gather witness statements and relevant environment conditions via notes, photos, and recordings. All data is streamed to a central database.

Safety managers use ArcGIS Dashboards to analyze escalator-related accidents, such as creating heat maps to identify high-risk areas in the airport. The dashboard shows the numbers and types of injuries for each escalator and ratios of types of escalator-related

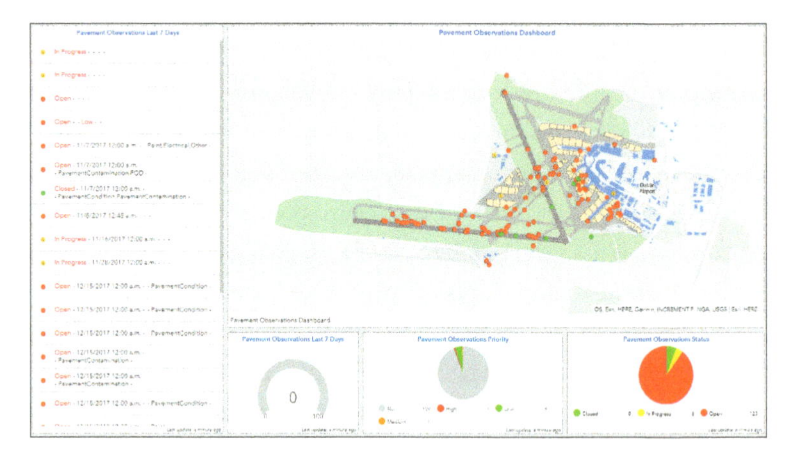

Dublin Airport uses ArcGIS Dashboards to show where there are currently problems with the runway pavement that need to be addressed.

injuries, such as collisions and cuts. GIS analysis highlights safety improvement opportunities and produces reports that protect the airport against erroneous claims. GIS analysis helps demonstrate compliance with European Union Aviation Safety Agency aerodrome licensing regulations by showing that asset management standards have been met.

GIS is used for other operations, such as maintaining the airfield. For example, inspectors use ArcGIS Collector to collect information, including photographs, on the condition of the pavement on the airfield. The data is displayed and prioritized in a GIS dashboard for reporting pavement conditions.

Additional GIS dashboards are used to monitor safety issues, such as maps and statistics that show patterns and instances when birds strike aircraft, helping staff with wildlife management plans.

This story originally appeared as "The Digitally Intelligent Airport" by Barbara Shields and Carla Wheeler, in *ArcWatch* (2017). All images courtesy of Dublin Airport.

TAKEAWAYS

TRANSPORTATION AGENCIES TYPICALLY HANDLE MANY types of operational situations and issues simultaneously, including repairs and maintenance and contending with the smooth flow of people, goods, and services. A common operational problem in any transportation system is that issues, such as repair work or an accident in one location, can have negative impacts on other parts of the system, affecting safety, causing delays, increasing operational costs, and even result in a decrease in customer satisfaction. With GIS, transportation agencies gain a geographic perspective of the entire system, bringing all the moving parts together to a singular system of understanding. Transportation agencies and authorities use GIS to create operational views of activities, such as maintenance, and establish an overall situational awareness of combined activities taking place across the business.

Key takeaways for applying GIS to transportation operational efficiency

- **Workforce collaboration:** Mobile GIS helps transportation agencies move away from traditional paper workflows, and plan and coordinate work activities in real time. Mobile apps help maintenance crews navigate to the next priority job site, access documents and schematics on site, and immediately update business systems and databases with the most current information.
- **Risk assessment:** Using GIS analytics, engineers, planners, and operations center managers can uncover potential vulnerabilities in transportation systems and visualize patterns and trends of historical data. Spatial analysis

allows agencies to identify and prepare for potential hazards, strengthen risk mitigation plans, allocate resources, and better understand impacts to assets, people, and places.

- **Performance monitoring:** GIS adds a higher level of location intelligence to assess how transportation systems perform over time and in real time. Performance metrics can be presented and interrogated at various scales, providing insights to systemwide operations and by specific locations. Location-based performance review of maintenance work, ongoing or completed projects, new construction, vendor leasing programs, and the movement of people, cargo, vehicles, aircraft, and vessels are common ways transportation agencies improve operational efficiency.
- **Data consolidation:** GIS connects location data and enables spatial analytics within other business systems used by transportation organizations. Interactive maps provide another way to assess and understand corporate data stored in asset management systems, computer-aided dispatch systems, document management systems, and customer relationship management systems, to name a few. GIS also combines data feeds from weather services, social media, and other data services to provide a consolidated view of the operational environment.

Learn more about applying GIS to transportation operational efficiency in the last section of this book, "Next Steps."

ROAD
CLOSED

PART 2

SAFETY AND SECURITY

NO MATTER WHO THE USERS ARE OR WHAT IS BEING transported or which transportation mode is in question, you can be sure that mountains of rules, regulations, protocols, and security measures are in place, designed to keep people safe and protect the infrastructure of the operation. And yet, transportation safety and security concerns are a never-ending list of what-if scenarios for both public and private transporters. A single incident or accident or security breach can quickly cascade into bigger issues, causing delays, disrupting business, cutting into revenue, and even impacting customer behavior and loyalty. Many transportation agencies, authorities, and businesses are using GIS not only to prevent safety and security incidents but to respond to those occurrences faster and more efficiently by placing them within a locational context.

Situational awareness and intelligence, that ability to evaluate and respond to issues in relation to the entire operational environment, is a key factor for solving safety and security problems. GIS allows organizations to create a collaborative system of understanding and coordination through continuous gathering and sharing of information.

To achieve a higher level of safety and security using GIS, transportation agencies typically begin by asking the following questions.

How do we create a comprehensive view of safety and security?

GIS gives transportation professionals the ability to integrate companion technologies, including surveillance feeds and sensor data, into a universal geospatial system that includes maps, GIS dashboards, and predictive spatial analytics for overall situational intelligence. Everyone—including workers on the ground, operations managers, executives, and partnering agencies such as local fire and police departments—can partake in and contribute to the system.

How do we synchronize incident management activities?

When something happens, such as an accident, a security alert, or a weather-related incident, transportation personnel are mobilized and work together to address the problem. With GIS, data is captured, processes are documented, and the status of incidences is monitored within a common operational environment—ensuring that everyone is working with the same information. GIS maps, real-time dashboards, and mobile apps are all interconnected, allowing responders and incident managers to prioritize, track, and report on multiple situations.

How can we empower and coordinate our mobile workforces?

Safety and security operations almost always require people on scene to manage incidents and fix the problem. These on-the-ground personnel can use GIS to provide supplemental intelligence about a situation, including photographs, video, and precise locations of relevant issues, to the command center. Likewise, command center managers can use GIS to retask responders or escalate response efforts based on new information from on-scene personnel while maintaining an overall operational picture of every work crew, vehicle, and asset.

In the following selection of case studies, you will learn how a major port created a GIS-based application that maintains a multilayered physical security system including maps and dashboards connected with closed-circuit television surveillance, radar tracking, sonar, and other sensor-based systems. You will see how a regional transportation authority used GIS to prepare for increased transit ridership after the coronavirus disease 2019 (COVID-19) pandemic as people started to return to their workplaces. In another story, a transportation department applied GIS to create a new, automated predictive road safety risk assessment methodology to make the national transportation infrastructure system safer. And finally, you will learn how a busy international airport integrated web and mobile GIS apps with its incident management procedures to plan for and respond to incidents.

STRENGTHENING PORT SECURITY

Port of Long Beach, California

THE PORTS OF LONG BEACH AND LOS ANGELES FORM THE San Pedro Bay Port Complex. The two facilities make up the largest port complex in the United States and are responsible for more than 40 percent of the nation's containerized cargo shipments.

Occupying more than 3,200 acres (or 13 square kilometers) of land with 25 miles of the waterfront, the Port of Long Beach poses a major challenge for security operations, particularly because it is an open port that provides docking services to pleasure and small-business craft as well as commercial cargo ships. In addition, more than 15,000 trucks and 100 trains move in and out of the port every day. The dynamic nature of the port, with its constant movement and 24-hour operations, requires close surveillance.

Like many ports, the Port of Long Beach maintains a multilayered physical security system that includes closed-circuit television surveillance, radar tracking, sonar, and other sensor-based systems, which are folded into the Virtual Port application.

"Virtual Port is an ArcGIS-based system that is fundamental to our security operations," said Randy Parsons, director of security for the Port of Long Beach. "It is the essential technology of our command center and has allowed us to geospatially enable our entire security operation. So, we now have more than 60 geographically referenced databases that are integrated with the existing elements of our physical security system, which increases exponentially our ability to monitor and analyze our daily operations."

The Virtual Port application is also used for two other primary functions: incident response and business recovery. "The beauty of Virtual Port is that we are using the same ArcGIS system with the same databases for all three of our main functions," Parsons said.

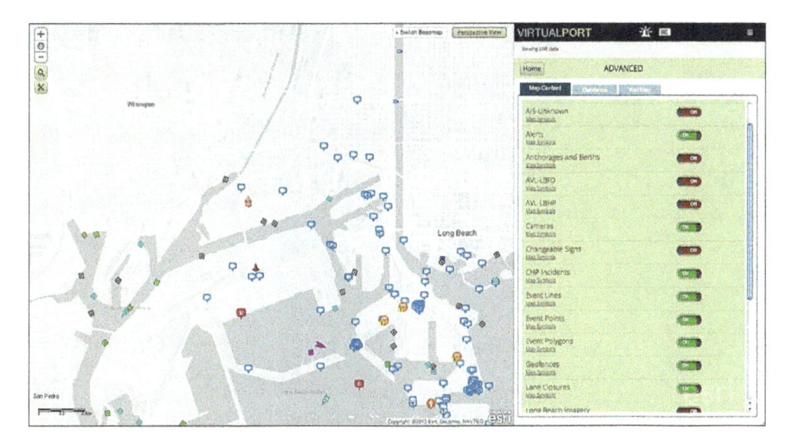

Virtual Port is a dynamic, ArcGIS-based system that is fundamental to security operations at the Port of Long Beach.

"This means that our security systems operators have complete familiarity with the system and don't have to switch over to something else in the event of an emergency."

The business recovery aspect of Virtual Port is to help officials determine which areas of the port are inoperable, what remains open, and what resources are available to quickly resume port operations. Partner agencies, such as the 12 law enforcement agencies that have personnel stationed at the port, can connect to Virtual Port and access its common operational picture to collaborate and share information, ensuring the resiliency of the facility. Port officials can also run what-if scenarios that model chemical plumes and other hazards to help agencies prepare for and better understand the impact of potentially dangerous situations.

Additionally, federal, state, and local agencies use some aspect of Virtual Port, including the City of Long Beach Emergency Communications and Operations Center, local police and fire departments, and local health agencies.

Besides security, Virtual Port is used for two other primary functions: incident response and business recovery.

"While Virtual Port provides us with a clear operational picture of our extensive day-to-day security activities, we have found that it is also helping lower our business operating costs by streamlining those processes," Parsons concluded. "This is a huge benefit to centralizing our security operations around ArcGIS because it provides us with an increasing return on investment."

This story originally appeared as "GIS-Based Security System Benefits Port in Many Ways" in *ArcUser* (Winter 2016). All images courtesy of the Port of Long Beach.

PREPARING FOR TRANSIT RECOVERY

Regional Transportation Authority of Northeastern Illinois

DURING THE COVID-19 PANDEMIC LOCKDOWN, THE Transportation Authority of Northeastern Illinois (RTA) began to anticipate an increase of transit ridership as people started to return to their workplaces. The RTA worked with the Chicago Transit Authority (CTA), Metra, and Pace Suburban Bus (the service boards) to strategize communication with riders on the second-largest public transit system in the country.

The RTA is the financial, budgetary, and planning oversight agency for the three service boards in northeastern Illinois. The CTA operates a rapid transit and bus system serving the City of Chicago and 35 surrounding suburbs. Metra operates the region's commuter rail system that includes 11 routes. Pace operates the suburban bus system and regional Americans with Disabilities Act (ADA) paratransit services. Collectively before the COVID-19 pandemic, the service boards provided nearly 2 million rides each weekday in six counties with 7,200 transit route miles.

Early in the statewide lockdown, the RTA experienced a 76 percent decline in ridership relative to 2019, with Metra ridership down 95 percent, Pace down 63 percent, CTA rail down 85 percent, and CTA buses down 64 percent.

Throughout the pandemic, the RTA has been responsible for providing updates about how COVID-19 has affected ridership, service, and revenue for the region's public transportation system to a broad group of stakeholders and the public. Brad Thompson, manager of Data Services and Analytics at RTA, and Hersh Singh, principal analyst, developed a GIS dashboard to support decisions about planning projects, maintenance needs, revenue sources, and fund allocation. The dashboard is used to pull data from numerous internal and external sources and visualize it with maps, graphs, and other storytelling tools.

In response to COVID-19, the RTA team launched the COVID-19 Transit Dashboard to share daily and monthly data with public officials at the local and state level. "This is something that's really never been done before," Thompson said about the dashboard. "We've never had to respond to a disruption of this scale and duration so quickly. We needed to find a way to compile data from a variety of sources and report on it in an online format on a weekly basis."

During the pandemic, many of Chicago's transit-dependent residents and essential workers have continued to use public transportation to get to their jobs, grocery stores, doctor appointments, and other places. Thompson and his team used GIS to identify locations of people and businesses still in need of transportation. The team analyzed a variety of census data sources as well as the Center for Disease Control and Prevention's (CDC) Social Vulnerability Index (SVI) to map densities of at-risk populations.

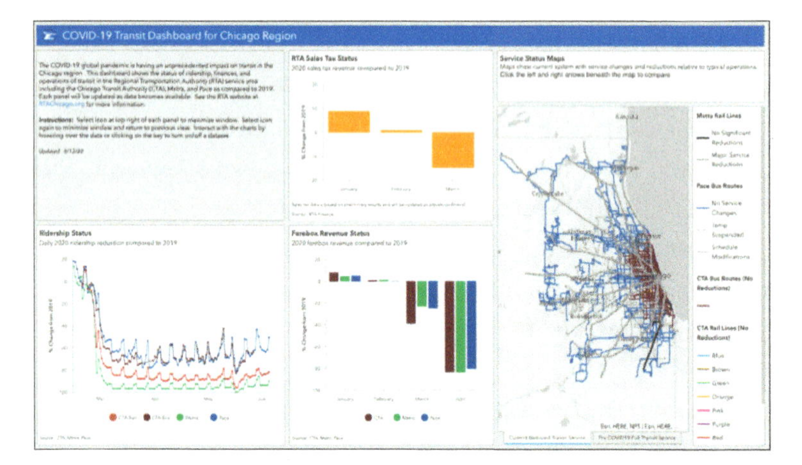

RTA's COVID-19 transit dashboard combines details about ridership, revenue, and service status. (This static screen capture is from July 7, 2020.)

"We created an internal mapping tool for RTA planning that geocoded all the supermarkets and hospitals in the Greater Chicago area to identify which of those were served by some type of public transportation," Thompson explained. "We also pulled in data identifying where essential workers were located throughout the region and compared it with the identified locations of essential businesses and RTA transit maps, which confirmed that there was a direct relationship between the commute routes of those essential workers and the public transit lines that were still operating and sustaining ridership levels."

When Susan Massel, director of Marketing and Communications at RTA, saw the COVID-19 dashboard, she realized it contained nearly all the information the agency had been including in its regular communications to elected officials and stakeholders in narrative form.

"The dashboard is a better way to share information," Massel said. "The tool was initially created to communicate with the public, but it's evolved into a great way to share information more broadly with a variety of audiences, including elected officials, stakeholders, and media."

This story originally appeared as "COVID-19: Mapping Chicago Region's Transit Recovery" by Terry Bills on the Esri Blog (July 21, 2020). Image courtesy of the Regional Transportation Authority of Northeastern Illinois.

ANALYZING ROAD SAFETY

New Zealand Transport Agency

THE NEW ZEALAND TRANSPORT AGENCY (NZTA) IS responsible for delivering a safe and efficient national transportation infrastructure system. NZTA developed a predictive road safety risk assessment methodology that specifically targets low-volume rural roads to evaluate the safety of curves based on a road's geometric and operating speed features. Esri partner Abley Transportation Consultants (ATC), which specializes in designing and implementing road safety strategies, helped NZTA build the risk methodology based on ArcGIS.

NZTA had two key requirements for the risk assessment methodology. First, the methodology had to be created using already available knowledge of the road network, including existing spatial data and associated transport attributes, such as speed limits. Second, it had to be cost-effective so it could be readily applied across an extensive road network.

ATC based the risk prediction methodology on a road engineering process that includes driver behavior models for acceleration on straight roads and deceleration on curves along high-speed corridors. The process involves comparing approach speeds with the radius, or tightness, of a curve to assign a risk classification to each curve in both directions of travel. This risk classification is a proxy, or model, for the likelihood that a driver will lose control while taking the curve.

Previously, a traffic engineer would split a road corridor into a series of straights (with known lengths) and curves (with known radii), and then follow a complex methodology to identify maximum desirable speeds based on the overall curvature and terrain profile of the road. Then, the transportation authority would categorize curves by comparing the approach speed to the curve radius. Although this method was effective in reducing crashes, manually implementing it across all the high-speed roads in New Zealand was too time-consuming and cost prohibitive.

For the predictive model, ATC developed several smart geospatial workflows to segment road corridors, identify curves, calculate curve radii, predict vehicle operating speeds along corridors, and assess curve risk based on approach speeds and curve radii across the entire high-speed road network. Using ArcGIS, the project team created a process for identifying curves and straights, splitting roads into 10-meter (32-foot) sections, and calculating the curve radius over a 30-meter (98-foot) arc. Individual curves were identified as contiguous sections with a radius of less than 500 meters (1,600 feet). ATC then modeled maximum desirable operating speeds for each section of road by combining the overall terrain (based on a digital elevation model) and curvature of the road, which ranged from flat and straight (allowing cars to go 110 kilometers per hour, or 68 miles per hour) to mountainous and tortuous curves (allowing for speeds of only 75 kilometers per hour, or 46 miles per hour).

From there, ATC calculated actual free-flow operating speeds and curve risk by running a Python script in ArcGIS that sequentially evaluated each element of the road—both curves and straights—to model vehicle speeds and driver behavior. On undivided roads, vehicle operating speed and curve risk is analyzed in each direction, making it possible to identify curves, where the risk to drivers occurs in only one or both directions of travel.

To validate the methodology, 10 years of crash data was compared for curves across the entire road network. The analysis revealed that two-thirds, or 67 percent, of loss-of-control crashes occurred on 20 percent of curves classified as unacceptable or undesirable in at least one direction. This finding suggests that by targeting a small

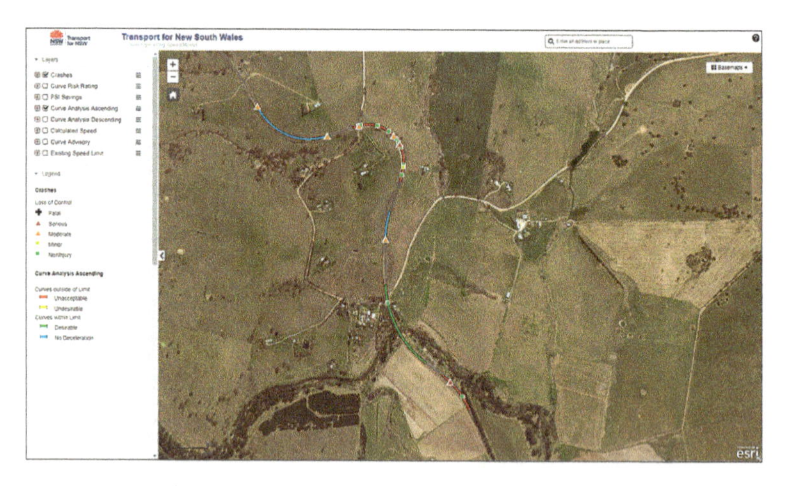

When curves require drivers to decelerate substantially, they can cause loss-of-control crashes, as the symbols on this map show.

percentage of high-risk curves for further investigation and intervention—such as improved signage, surfacing, or side protection—road agencies could greatly reduce the likelihood of additional crashes occurring at high-risk locations.

Since the predictive road safety risk assessment methodology assesses the risks independent of crash information, NZTA can now proactively target interventions toward high-risk locations and have confidence that safety will be improved—whether the location has an established crash problem. The methodology effectively bridges the gap between being aware of major safety issues on high-speed roads and developing detailed strategies that reduce the likelihood and consequences of crashes that happen on roadway curves.

"These guides and tools have revolutionized the way we and our partners are able to view and understand the various road safety risks across the New Zealand road network," said Colin Brodie, lead adviser of interventions for safety and the environment at NZTA.

The road safety risk assessment methodology has been rolled out across all high-speed roads in New Zealand, which consists of 42,000 kilometers (26,000 miles) of roadway. The Centre for Road Safety in New South Wales, Australia, has also implemented the methodology on 37,000 kilometers (23,000 miles) of state-owned roads.

This story originally appeared as "Smarter Data for Safer Roads" in *ArcNews* (Winter 2018). Image courtesy of the Department of Transport for New South Wales.

IMPROVING INCIDENT MANAGEMENT

Los Angeles International Airport

MORE THAN 70 MILLION PEOPLE PASS THROUGH LOS Angeles International Airport (LAX) in a year. Maintaining situational awareness to manage safety, security, and operations is complex. At LAX, the Airport Response Coordination Center (ARCC), its command and control facility, oversees field personnel who work shifts 24 hours a day, seven days a week, both airside (past security checkpoints) and landside (public facing).

To manage incidents and security, the ARCC staff integrate security and operational intelligence—including video surveillance and closed-circuit television—to generate automated adaptive response plans. Until recently, however, much of the collaboration between ARCC staff and LAX field personnel was done manually. Control room staff and mobile personnel communicated information and status updates via phone calls and two-way radios. Mobile personnel would call or radio in incidents, and an ARCC operations superintendent would then assign technicians via more phone calls or radio communications. All information was conveyed verbally, without maps. Airport officials decided to speed up communication by creating a common operational picture (COP) that control room staff and mobile personnel could all access, no matter where they were located.

LAX worked with Qognify, a video management software and enterprise incident management solutions provider, and technical and management support consultant AECOM (an Esri partner) to develop a solution that improved the airport's COP. The idea was to leverage LAX's existing incident management system by implementing ArcGIS Enterprise, which would allow maps and geographic information to be accessed anywhere, anytime, on any device. Together, Qognify, AECOM, Esri, and LAX created a program called the Qognify Situator eGIS web application. A two-part solution, the eGIS web application gives LAX's incident management efforts both spatial and web capabilities.

In addition to integrating mission-critical information, the incident management system also includes IBM Maximo software to manage work orders and AirIT's PROPworks to handle lease and property management. The new system gives users a complete picture of any situation, including airport buildings, property, infrastructure, and security sensors.

"You can't separate operational events and incidents from the infrastructure you're trying to protect or enhance," said Dom Nessi, former deputy executive director and chief information officer at Los Angeles World Airports.

ARCC staff access the data through both the Situator (a desktop map program) and the eGIS web application, which display geographic information in real time. Control room staff use the Situator

Accessing the eGIS Web Viewer on iPads, field staff can initiate incidents by marking them on an up-to-date map.

to manage incidents. Then they extend the control room to the mobile staff with the eGIS web application.

Using the eGIS Web Viewer on iPads, mobile staff can initiate incidents—a slip and fall, a fuel spill at a gate, a leaky toilet, or a pothole on the taxiway—by marking them on a map in the application. They can then add associated information, such as photos and an incident type.

Back in the control room, a collapsible panel next to the map tracks annotations from the field so that, as incidents unfold, ARCC staff can gain insight and engage in spatially smart dialogue with mobile crews. Comments are time-stamped to provide an audit trail of each event.

With the desktop and web applications, ARCC staff can visualize where they need to send personnel during an incident. If a situation involves travelers, staff can see how to evacuate them or at least get them out of impact zones. To make this work, control room staff share their own real-time map annotations with mobile crews.

The eGIS Web Viewer also integrates all predefined workflows from the Situator application, so users can pull up specific, preplanned procedures for event response. For an incident in a terminal, for example, mobile crews can see where to set up command posts, query nearby assets (such as the closest surveillance camera or an underground hydrant fueling line), draw routes, spatially scope accidents (such as fuel spills), and add notes. They can access lease information to view tenants in impact zones, their store hours, and contact information. As staff move through the steps to mitigate an occurrence, they can use the program to check off each component of the appropriate incident management procedure.

"Pulling that information out of the GIS and putting it into the hands of those who need it in real time was a big step," Nessi said. "You can say there is water leaking in Terminal 1, but until you have

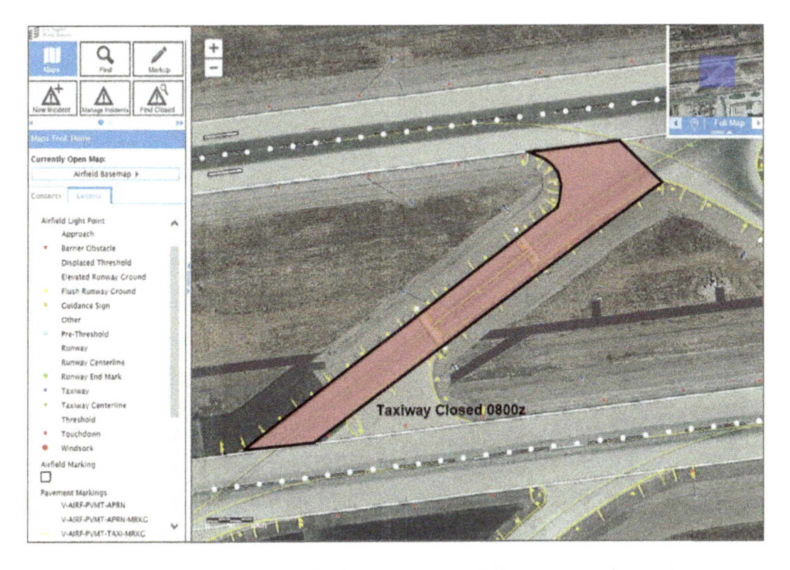

ARCC staff access data through the Situator and the eGIS web application, which both display visual geographic information in real time.

an exact picture of where the water is, you can't do anything. Now staff can not only report it, but they can take a photo of it and attach it to the work order so that people in the ARCC see exactly what they're seeing in the field."

More than 200 airport staff have access to the eGIS Web Viewer across security, operations, property management, information technology, and airport development groups. "They have everything they need to respond to an event at their hands," said Kevin Carlson, vice president of national aviation technology at AECOM.

The app brings daily operations into the airport's enterprise GIS. This includes normal activities such as examining pavement markings or safety areas, as well as emergency situations including wildlife strikes or malfunctioning baggage carts.

By using the eGIS web application and mobile app, LAX has enhanced compliance by better documenting event response. And having historical data records has enabled staff to spot trends, such as recurring electrical issues, which helps them implement longer-term solutions. But LAX's enterprise GIS is no longer just an archive of engineering data. Rather, it conveys critical, real-time information spatially, from the control center to the field and back.

"Through its deployment of the ... eGIS Web Viewer, LAX has really raised the bar in terms of empowering its personnel in their daily management of airport safety, security, and operations," said Qognify president Moti Shabtai. "People outside the control room can be part of [an] incident management process by initiating incidents from the field when they see something happening. They can get real-time geospatial information and receive adaptive response plans as well. All these capabilities enable LAX to contain incidents faster."

This story originally appeared as "Real-Time GIS Improves Incident Management at LAX" in *ArcNews* (Winter 2016). All images courtesy of Los Angeles International Airport.

TAKEAWAYS

GIS HAS BECOME A FUNDAMENTAL TOOL FOR IMPROVING transportation safety and security. Transportation agencies and companies use GIS to prepare for potential incidents and emergencies by identifying where vulnerabilities exist and improving communication between personnel and other agencies using maps and spatial analysis. When responding to emergencies, transportation security personnel use real-time GIS dashboards to support internal decision-making and coordinating with other groups, such as local police and fire. Once the incident or emergency is over, safety and security teams create recovery maps and share ongoing recovery efforts.

Key takeaways for applying GIS to transportation safety and security

- **Resilience:** GIS combines layers of data on critical infrastructure, natural hazards, and human-caused threats and on the surrounding community's demographics and social inequities to help locate where to invest resources to change the outcomes. Data-driven insights help transportation organizations prepare for worst-case scenarios and build programs that address the underlying vulnerabilities. Safety and security personnel use GIS tools for visualizing and analyzing risk, performing real-time analytics, managing mitigation projects, and engaging with the public.

- **Response:** With multiple complex emergencies that can exceed an organization's limited resources, transportation authorities use GIS to become more agile and informed at all points during response. With real-time dashboards for situational awareness, users can share information and context using maps, apps, and tools that allow personnel to quickly adjust response strategies and tactics based on rapidly changing conditions. Safety and security teams can maximize impact and reduce the time needed to respond to whatever comes next.

- **Recovery:** GIS helps safety and security professionals quickly assess damage, investigate causes, open roadways and facilities, and coordinate the removal of debris or dangerous material. Mobile GIS tools facilitate building a reporting database that is essential to ensure short-term recovery, estimate costs, and support funding requests. Where recovery is long term, understanding where to place safety and security resources, identifying future vulnerabilities, engaging local communities and other emergency responders, and communicating how the recovery is progressing are essential to restoring normalcy.

Learn more about applying GIS to transportation safety and security in the last section of this book, “Next Steps.”

PART 3

ASSET MANAGEMENT

THINK OF ASSETS AS ALL THE PIECES AND PARTS THAT A transportation organization manages and maintains. Assets can be tiny, such as a sensor on a road sign, or large, such as a building or property owned or leased by the organization. Keep in mind that assets are components of a bigger transportation system, and that the failure or replacement of an asset may affect other assets in the system and disrupt at least a portion of the organization's daily operations. For example, a conveyor belt for moving passenger luggage to the appropriate carousel inside a baggage claim area is a component of the baggage handling system used by many airports. If the conveyor belt breaks down, a part of the baggage handling system is affected, and a maintenance crew is deployed to fix the problem.

Every asset has a location within the broader system. Most transportation organizations use asset management software and databases to keep an inventory of assets. An asset management system manages the entire life cycle of physical assets—from design to disposal. When GIS and asset management systems are combined, the distributed and interconnected nature of asset locations becomes apparent, helping the organization process, synchronize, and share information in an efficient and effective manner across departments. The combination of GIS with an asset management system is referred to as enterprise asset management (EAM), in which GIS is used to analyze and display the spatial relationships between assets and other aspects of the business over time and geography. For example, a road and highway department can use an EAM to map electronic signage repair history and correlate failures to a specific component in the signage during severe weather conditions, and then plan and monitor the work of replacing that component across the entire system.

Outside of construction, asset management may represent the largest expense in any existing transportation system. The application of GIS to asset management provides an additional level of knowledge—called location intelligence—that transportation agencies can use to better understand where and why problems are happening.

Adopting location intelligence within asset management begins with asking a new set of questions.

How do we identify and visualize the location of every asset?

A prerequisite for good asset management is a comprehensive asset inventory that includes identifying the precise locations of all assets—from equipment to infrastructure to property. GIS provides the capability to accurately collect asset locations; organize asset data into spatial views, such as maps and GIS dashboards; and access existing asset data, documents, and photos within web and mobile applications.

How do we use location to coordinate asset inspection and maintenance work?

Effective asset management requires that everyone in the organization, from the asset manager to the onsite repair crew, has access to the same information to properly conduct daily tasks, including asset inspections and maintenance work. Location ties the work and data together. Maintenance crews and inspectors use GIS to capture fresh, detailed information that automatically updates corporate databases and EAM systems, helping to document work, prioritize work orders, and retask crews based on their locations and proximity to other issues and assets.

How do we gain a systemwide spatial view of asset performance?

GIS helps asset managers maintain a state of good repair and improve the efficiency of maintenance activities and allows them to move from reactive maintenance to predictive maintenance—anticipating problems and avoiding costly asset failures. Asset managers can use GIS analytics and dashboards to monitor key metrics that highlight performance, present comprehensive data in a clear and concise manner, and allow managers to focus on what is important and keep operations running.

In the following selection of case studies, you will learn how the largest transit agency in the US uses GIS to better coordinate asset repair work, including where and how it was performed. You will see how a department of transportation transformed its statewide data governance policies and shifted from old data entry processes to modern mobile GIS data collection, which allows it to share asset data throughout the organization. In another story, one of the busiest ports in the world overcomes limited space and expands capacity using a modern GIS solution to untangle disjointed legacy systems that restricted access to asset data. Finally, another state department of transportation uses GIS to improve roadway safety, maintenance, and longevity by retooling its strategic performance initiative with location intelligence.

OVERSEEING $1 TRILLION IN ASSETS

New York Metropolitan Transit Authority

NEW YORK CITY MANAGES THE LARGEST TRANSIT SYSTEM IN the United States, servicing nearly 9 million bus and train riders every weekday. Stretching through New York City's five boroughs and the surrounding suburbs, the transportation system encompasses hundreds of miles of track, more than 12,000 buses and train cars, and a staggering $1 trillion in hard assets. These assets range from the signals embedded in tracks to the tracks themselves, along with the bridges and tunnels that form the system's arteries.

When Sean Fitzpatrick, director of enterprise asset management for New York's Metropolitan Transit Authority (MTA) discusses the challenges facing the New York subway, he begins with a simple concept. "There's a huge onus on the MTA to run 24/7," Fitzpatrick explained. (If subway riders could chime in, they might add, "and run flawlessly.")

MTA had been working for years toward better asset management. In one of the world's largest subway systems, it's no small feat to understand where every asset is located, let alone keep each one in working order to ensure that millions of New Yorkers can crisscross the city every day. MTA's allotment is approximately $30 billion.

"Even though our budget sounds big, when you're talking about $1 trillion in assets, it's limited in what it can accomplish in a system as large and complex as ours," he said. "We have to make sure we're spending money in the right way."

Previously, when assets broke down, the MTA's approach was to assemble a crew of trade workers including:

- A safety team to monitor the work and communicate with a subway control center

- Signal electrical engineers
- Track mechanical engineers
- Track electrical engineers

The workers were costly and often difficult to coordinate. The break-and-fix routine was papering over the system's ills and draining the MTA's budget.

"When you fix something after it has broken, it's three to four times more expensive as opposed to having done something either preventatively or predictably," Fitzpatrick noted. "So, we're on a journey to move from reactive to preventative to predictive, ultimately."

As part of that shift, the MTA committed to mustering work teams more efficiently, deploying them to work sites with precision, and helping them accomplish more while they're together. To do that, managers and work crews needed location intelligence. "One of the things that we did almost immediately was, we started to use GIS technology to map where defects were occurring, both in the signals and on the track side," Fitzpatrick said.

For Fitzpatrick and the MTA, the location intelligence gained from GIS revealed opportunities for efficiency that they had been missing. "We were able to see very quickly that there were certain signals and certain sections of track where we were having recurring issues within a very short period of time—days [or] weeks where things were repeatedly breaking down," Fitzpatrick said.

Once the MTA mapped defects in its signal system, managers moved to better coordinate the repair work itself, including where and how it was performed. They created a GIS-based mobile app for joint switch inspections—the jobs that require teams of skilled craft people. Now when a manager requests a signal repair, all members of the crew receive a work order on their mobile device. Through an integration with the MTA's logistics system, it's clear when the needed materials will arrive on the job site.

With the mobile app, the MTA has also begun to shift from reactive to proactive asset management. For example, a crew manager can use GIS capabilities to find other signals and switches in the vicinity of a scheduled repair. If a similar switch is nearby, the team can perform preventative maintenance on that asset as well. The app has reduced the time required for joint switch inspections by nearly 50 percent, resulting in significant cost savings and improved subway uptime.

Although location intelligence technology helps the MTA log short-term improvements, it's also fueling long-term plans. The agency is using drones to perform inspections of aboveground rails. Underground, track-geometry cars roll through the tunnels, outfitted with sensors that collect data on track curvature, temperature, moisture, and other physical conditions. The sensor data is fed into the GIS, which logs the location of those measurements, helping managers identify anomalies and equipping passenger trains with those same sensors for more frequent inspections.

"Feeding that [data] into GIS, we can then look at those [assets] spatially to figure out not just the condition of the track, but are there certain curves where we experience more issues than others?" Fitzpatrick said. Safety inspectors will analyze, for instance, whether certain levels of wear boost the likelihood of derailment on a particular section of track and whether a slight change in the curvature might reduce that risk.

Moving forward, the MTA wants to use the same geospatial technology to create a digital twin of its assets—a virtual 3D map showing infrastructure locations and conditions—projected onto mobile device screens or augmented reality headsets, allowing MTA work crews to virtually see through the pavement for a bird's-eye view of tunnels, tracks, signals, and utility infrastructure affecting the maintenance work they need to perform.

This story originally appeared as "How the New York MTA Manages $1 Trillion in Assets" by David LaShell in *WhereNext*, August 28, 2018.

ADVANCING DATA GOVERNANCE

Florida Department of Transportation

TRADITIONAL TRANSPORTATION NETWORKS AND accessible technology now have evolved to become intertwined and inseparable. As technology continues to advance and transportation systems begin to bring in more big data, departments of transportation need to assess—and potentially change—the cultures they have developed around technology.

At the Florida Department of Transportation (FDOT), we have examined the way we collect, govern, and use data and technology in our day-to-day operations and business obligations. Now, instead of trying to build our way out of challenges—which can be slow, expensive, and fleeting when talking about physical infrastructure—technology is a core agency strategy. This means that technology, especially GIS, always has a seat at the table at FDOT and is never an afterthought.

FDOT is responsible for one of the largest and most extensive transportation systems in the country. Supporting more than 20 million residents and over 100 million visitors annually, Florida's transportation network includes 15 seaports and 2,890 miles of navigable waterways, 2,895 miles of rail line, 122,659 centerline miles of public roadways with 12,262 bridges, 779 airports, 2 spaceports with 10 launch facilities, and 31 urban transit systems.

With assistance from developing technologies, each aspect of Florida's transportation system produces data—and a lot of it. How to best use and manage this growing amount of data became an important point when discussing the intersection of transportation and technology. But FDOT had always assessed, planned, and financed transportation and internal technology initiatives separately. Because of this, transportation technology was not being funded appropriately, even though the nature of transportation development was inherently growing more technology oriented. This disconnect placed the agency responsible for shaping the transportation of the nation's fourth-largest economy at a serious disadvantage and a risk of falling behind in terms of growth and innovation.

To meet the challenges that both big data and emerging technology were posing to Florida's transportation systems, FDOT established its aptly named ROADS initiative, which stands for Reliable, Organized, and Accurate Data Sharing, to create and implement an integrated enterprise information management system.

The agencywide movement set out to ensure that FDOT's current and future data is accurate, secure, and reliable in the hopes that it can empower employees to perform their jobs at an even higher caliber. With improved internal communication, data collection, and storage, FDOT's employees and stakeholders can more efficiently and reliably access relevant business data and share it across the department.

Around the same time that we launched the ROADS initiative, the Federal Highway Administration (FHWA) had also begun issuing national guidelines and standards on data governance to all state transportation entities. The aim was for departments of transportation to organize and implement an institutionalized set of policies, procedures, structures, roles, and responsibilities when it came to managing data and information. From these guidelines, we explored the concept of civil integrated management (CIM) in relation to FDOT's own business practices. With CIM, the idea is to collect, organize, and manage data related to highways, bridges, and other transportation assets using a single, authoritative source that offers standardization and easy access to users.

Alongside FDOT's ROADS initiative and through the conceptual exploration of CIM, the agency also established the Office of

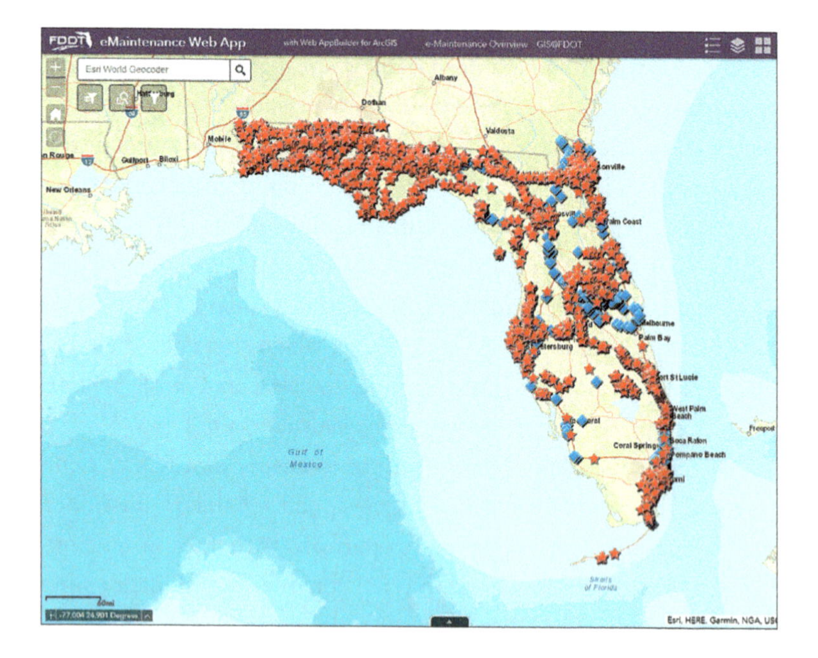

With the Florida Department of Transportation public-facing eMaintenance web app, anyone can see inspection results for crash cushions and guardrails across the state of Florida.

Transportation Technology (OTT). This formally solidified the relationship between key transportation technologies and the funding for transportation infrastructure. OTT required an internal reorganization that resulted in the newly established office that houses FDOT's Information Technology division and the Office of Civil Integrated Management, named for and focused on the same data governance principles implemented by FHWA.

"CIM is about leveraging advanced analytic tools with our existing transportation assets to do a better job in forecasting trends, conditions, and remediation," said John Krause, CIM officer and ROADS strategic liaison at FDOT. "It will provide us a safer, more cost-effective, and ultimately better transportation network for all Floridians, businesses, and visitors."

Our implementation of CIM included getting an independent assessment of FDOT's information technology capabilities, personnel, and infrastructure. The critical evaluation identified that we had a fundamental disconnect between transportation initiatives and technology plans. Like the allocated funding, the strategic plans for transportation business were developed and considered separately from the agency's technology business, hindering our potential and limiting how well FDOT could prepare for the cultural changes that were already occurring and yet to come.

With the implementation of ROADS, CIM, and OTT, FDOT started integrating technology initiatives and transportation planning in a way that is organically tied to its daily business processes. Suddenly, technology was no longer a footnote. And with the aid of GIS, we have built a new culture at FDOT for how we integrate technology with transportation assessment and planning.

The department's Office of Maintenance provides a prime example of how this transformation has gone. Until recently, FDOT employees used clipboards, paper, and pens to record the condition of various transportation facilities in the field. The physical reports then had to go through a lengthy data entry process before being collated into reports required by FHWA. But with mobile GIS technology, such as ArcGIS® Collector, employees can now record all necessary information digitally, on the spot, and share the data with others throughout the organization. The director of the maintenance office can now compile FHWA-required annual reports in minutes.

The changes we have made to data management and governance at FDOT—which we maximize by using GIS—save time and money. They are also making it possible for us to sustain this increased efficiency and greater transparency. For example, we implemented

ArcGIS to support the agency's electronic document management system, and now employees can easily combine business plans and related documents (design plans, contract information, project financials, and more) to create visual presentations. Decision-makers throughout the department use these resources to guide teams and assist with planning. And the success of our technological initiatives has driven demand for accessible and transparent data solutions at all levels of the department. As more people have started using GIS over the last few years, the number of ArcGIS® Online users at FDOT has quadrupled.

None of this would be a reality, however, if we hadn't implemented a new vision for data governance. Now, the CIM concepts adopted by the agency, together with the increased integration of GIS, have revolutionized the way FDOT does business. Because we have enhanced the relationship between technology and transportation from an institutional perspective, we now have room to innovate and are prepared to handle the big data that is starting—and will only continue—to come down the pipeline.

This story originally appeared as "Big Data Is Coming, and FDOT Is Prepared" by April Blackburn and Lydia M. Mansfield in *ArcNews* (Winter 2018). Image courtesy of the Florida Department of Transportation.

EXPANDING PORT CAPACITY

Port of Rotterdam, Netherlands

HOW DOES THE LARGEST PORT IN EUROPE GROW ITS business capacity when it can't expand its physical footprint? Is it possible to double operations without the ability to gain more space? Those were the questions the Port of Rotterdam confronted. Hemmed in by cities and water on all sides, the port was forced to take stock of its operations, its goals, and its strengths and weaknesses.

The Port of Rotterdam, established in the 14th century, is a marvel of engineering and transportation. It covers almost 30 acres of land and extends 26 miles inland from the water. It's the eighth-largest port in the world and the largest port in Europe—a gateway to 500 million European consumers.

Today, the port operates around the clock 365 days a year. In a typical year, 35,000 ships (nearly 100 a day) visit, carrying 400 million tons of cargo; 80,000 barges enter it; 7.5 million trucks traverse its roadways (that's 25,000 trucks every day); and 80,000 employees come and go for work.

In 2013, the port set a goal to grow from 400 million tons of cargo per year to 750 million by 2030. Allard Castelein, the port's chief executive, put it succinctly: "The port should become faster, smarter, and more sustainable."

But from a business process standpoint, the Port of Rotterdam was a tangle of disjointed legacy systems that put employees and assets in silos. Port leadership began to examine what needed to change. "We had to let go of everything. We went back to the start and asked ourselves basic questions: What are we? What is a port?" said Erwin Rademaker, port manager.

Ultimately, the solution was clear: "The only thing that is left for us to do is to improve, or to optimize, what we have," explained Rademaker. But the billion-dollar question remained—how?

At the time, port developers, business managers, project managers, asset managers, environmental advisers, port harbor operators, financial analysts, and many others made the daily decisions that kept the port running. Each group used its own system for data collection and reports; even more confusing, groups used different definitions for the same terms across the port.

The port, its leadership realized, needed one authoritative source of information for all users and assets—a single point of entry that would allow anyone, anywhere, to quickly access the data they needed to make smart decisions and perform their jobs more efficiently.

"We needed [another] modality in our port. And that is information. Because nothing in the port moves without information.

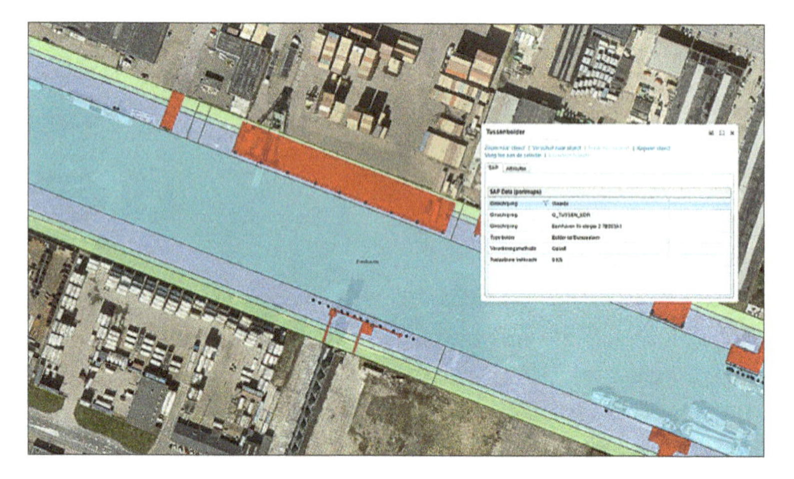

In the new digital viewer, users can select any area in the port and access live information from SAP and the document management system.

Everything in the port, from the largest berth to the smallest lock and key, is connected by information," said Rademaker.

After researching best-in-class information systems, the Port of Rotterdam set its sights on creating a new solution based on GIS technology. At its core, the solution is a beautifully simple map of the port. Underneath the skin of the map lie terabytes of big data—all accessible within three mouse clicks—and with connections to SAP, Microsoft Office, and a document management system.

The port recognized that the port and its users comprise three distinct spatial components, and that each one drives revenue generation in a different manner: (1) the land component is used by terminals and infrastructure, (2) the water component is used for transportation, and (3) the interface between land and water is used for the mooring of ships. With these simple divisions, it was straightforward to assign existing layers of data to the 10 core objects that make up the new port data system.

During implementation of the new GIS remedy, port leaders managed to phase out 49 other systems with relatively little disruption. Employees participated in the data migration, making the process one of active learning and training. And teenage children of employees tested the new interface, making sure it was user-friendly.

Today, more than 1,000 digital maps are created each day to guide operations and decision-making. In fact, all Port of Rotterdam data is presented visually. Any employee can pull up a map on a computer or mobile device, navigate to an area of the port, and click for more information. For instance, clicking on a wharf shows maintenance information, current contracts, ship movement data, and more.

Employees can also generate maps based on their specific needs. Rather than working with spreadsheets and lists, a business manager can pull up a lease expiration map and quickly see which areas are occupied, under reservation, or free and view the details of existing contracts.

Since implementing the new system, the port has seen an increase in throughput to 461 million tons, a 15 percent increase since 2013 and a significant step toward its 2030 goal.

Aside from financial goals, the port sees its new GIS-based approach as a means to become a world-class port. "Instead of being the biggest port in the world, which we were for decades," said Rademaker, the port manager, "we want to be the best port in the world. That means the most responsive to our customer needs."

This story originally appeared as "Intelligent Growth: The Digitalization of Europe's Largest Port" on esri.com, 2019. Image courtesy of the Port of Rotterdam.

EMPHASIZING PERFORMANCE

Iowa Department of Transportation

DEPARTMENTS OF TRANSPORTATION (DOTS) ACROSS THE United States are shifting the way they operate and manage assets in response to autonomous vehicles, rideshare services, and feedback from sensor data. "Are we in the business of facilitating mobility or providing the expertise to build highways and transportation structures?" said John Selmer, director of the Strategic Performance Division at Iowa DOT. "Traditionally the emphasis has been on the latter, but I think that's changing."

Iowa's DOT is no stranger to transformation. The agency took the lead in 2012 when a change in executive management and a federal law called Moving Ahead for Progress in the 21st Century (MAP-21) placed an emphasis on performance and outcome. The law required states to show the impact of their investments in infrastructure condition, safety, and system reliability (including metrics on congestion and travel time). Iowa DOT looked at the guidance and retooled, setting out to instill a measurement-oriented culture. A first step was the creation of the Strategic Performance Division that pulled experts together from different divisions.

"A hierarchical structure makes it hard to drive change because it's built for sustainability, consistency of product, and risk mitigation," Selmer said. "We work on cross-departmental projects such as asset management, data governance, business analytics, and project management."

The Iowa DOT had already integrated its location intelligence platform using GIS with other enterprise systems to ensure that data is maintained centrally in a GIS hub and shared with other systems as location as a service. "Creating the central GIS hub extended our reach into all other major business systems," said Eric Abrams, GIS manager at Iowa DOT. "It allowed us to evolve."

The Strategic Performance Division uses the GIS hub to tap into data from the other business systems, giving the means to quantify DOT overall performance and across divisions and field districts. "A lot of people within business units look at data as the dessert or the salad," Selmer said. "I look at data as the plate. Data is foundational, and it becomes exponentially more valuable when you combine it with other data."

Iowa DOT now embraces open data, assuming that all data should be open unless proven otherwise. This philosophy makes the department nimble, able to combine key information within purpose-driven applications to address business challenges.

Iowa DOT's Office of Maintenance took advantage of its access to data by creating a simple dashboard to measure the amount of salt

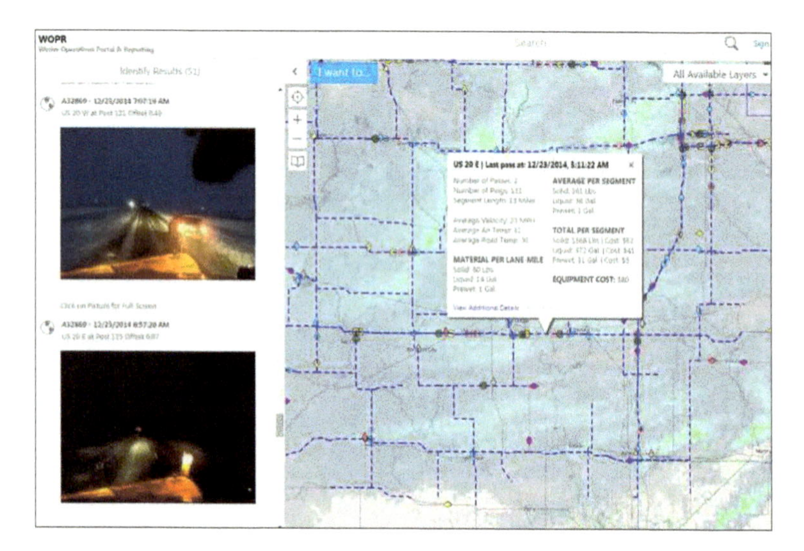

When the snow is falling, citizens have a one-stop place to see the conditions in their area and across the state.

applied to icy roads each winter. The salt measurement dashboard app showed DOT maintenance supervisors and other managers the quantity and cost of salt compared with previous years. Iowa DOT credits this simple comparison dashboard with saving the department $2.7 million a year within its first few years.

The Strategic Performance Division also used data from sensors aboard snowplows and along the road network to create a live map of winter driving conditions. The Iowa Track-a-Plow map was one of the first to offer this level of awareness, and it became an immediate hit with the public. "There were concerns with more eyes viewing our work in a more intimate way," Selmer said. "We have always felt responsible for the public's safety because we're the ones out there clearing roads. Ultimately the public plays a larger role in road safety, and the real-time understanding of conditions helps individuals and businesses make better driving choices."

With the data from Track-a-Plow, Iowa DOT created another dashboard view of additional metrics related to cost and performance. The statewide Winter Cost Calculator dashboard takes all data collected from individual plows—along with the cost of material, labor, and equipment—and aggregates it for each segment of the road network. The information is posted on an interactive map for DOT management and the public, putting the details from each storm and each year's operations alongside historical records for perspective. "Giving the public an understanding of the cost of a storm allows us to have more meaningful discussions," Selmer said. "We can create a better dialogue about levels of service and what it takes to keep the roads clear."

A great deal of work goes into building quality pavement. Smoothness, density, and physical appearance all factor into roadway safety, maintenance, and longevity. Iowa DOT continues to push the use of GIS into the field, recently adding an application to allow construction inspectors to enter pavement inspection data on tablets rather than fill out paper forms. Iowa DOT equips 157 full-time inspectors and 150 maintenance employees with tablets to view and collect data. They use one simple application to report inspections, reducing time spent from 30 minutes to just five minutes.

A digital workflow saves Iowa DOT hundreds of hours every year and makes it possible to immediately share data from across the state. Managers can view data from each pavement project. If a problem arises with a certain type of material, they can pinpoint where that material has been used. This added awareness increases the awareness of field crews and provides a new level of pavement asset management.

After taking steps to gain better insight into its own operations, Iowa DOT looked at the role transportation plays in a strong economy. The department hired a supply chain consultant to conduct research about the shipment of goods in and out of the state. The analysis led to a few surprises. It showed a great deal of grain traveling north to bakeries in Minnesota rather than out of the country. Analysis also led to a reconsideration of fleet efficiency as the Iowa DOT learned that, on average, only 60 percent of the available truck capacity is used. "The analysis showed how our system is being utilized," Selmer said. "We're looking at where we might place cross-docking and transloading sites so that truckers can combine loads and add capacity. That would make it much more cost-efficient for the shipper, and we'd have less trucks on the road."

In another effort, Iowa created a testing zone for self-driving fleet vehicles on a 20-mile stretch of Interstate 380 between Cedar Rapids and Iowa City. The state is using the exercise to gain a greater understanding of the data-harvesting implications from the large amount of data these vehicles generate. Iowa is also exploring the insights that can be gained through analytics. "Our initial guiding principle was really to get out of the mud and build infrastructure," Selmer said. "Now, we're looking at ourselves more as a mobility service provider, starting with a close look at how we operate on a 24/7 basis."

This story originally appeared as "Iowa DOT Makes Foundational Changes to Modernize Mobility" by Terry Bills on the Esri Blog, May 24, 2018. Image courtesy of the Iowa Department of Transportation.

TAKEAWAYS

GIS ADDS LOCATION INTELLIGENCE TO TRANSPORTATION asset management systems. With GIS, transportation agencies and companies can create a comprehensive asset inventory that includes the precise locations of all assets. Maintenance crews and asset inspectors can use GIS to capture detailed information that automatically updates corporate databases and EAM systems. Asset managers can document work, prioritize work orders, and retask crews based on their locations and proximity to other issues and assets, and move from reactive to predictive maintenance.

Key takeaways for applying GIS to transportation asset management

- **Asset inventory:** Transportation agencies use GIS to identify the precise locations of all assets and create a comprehensive, spatially enabled asset inventory that integrates with an existing asset management system.
- **Asset data collection:** Transportation asset inspectors and maintenance crews use GIS mobile apps to accurately collect new asset locations, update existing inventory, and monitor maintenance activities.
- **Performance monitoring:** Transportation asset managers use GIS to track asset condition and performance in real time throughout the life cycle, minimizing risks and maximizing performance.
- **Modern workflows:** Transportation asset managers use GIS to improve coordination and operational efficiency

in workforce activities and replace legacy systems with modernized workflows and reliable and accessible data.

- **Compliance:** Transportation agencies use GIS to comply with national guidelines and standards on data governance and to improve policies, procedures, roles, and responsibilities for managing asset data and other information.

Learn more about applying GIS to transportation asset management in the last section of this book, "Next Steps."

St. John's Wood
Baker Street
Great Portland Street
Euston
Euston Square
Warren Street
Regent's Park
Goodge Street
Oxford Circus
Tottenham Court Road
Bond Street
Marble Arch
Green Park
Piccadilly Circus
St. James's Park
Sloane Square
Westminster
Victoria

PART 4

PLANNING AND SUSTAINABILITY

BUILDING AND MAINTAINING TRANSPORTATION infrastructure represents enormous investments in time and money. To effectively plan investments, transportation authorities and agencies must analyze and evaluate projects, taking into consideration the long-term sustainability and resiliency of the infrastructure while projecting anticipated growth in business or demand and the changing needs of customers. GIS mapping and analysis provides a unique geographic perspective to understand current conditions and existing stresses on transportation systems, including demographic and lifestyle data to understand the current and future needs of customers and businesses, economic development patterns, and how to meet state and federal requirements.

The goals of transportation agencies are varied but most consider common themes. Take for example the San Diego Association of Governments (SANDAG) vision for transportation in Southern California. Key strategies include plans to:

- Enable new and better services for residents, transportation operators, and planners through technology
- Provide safe and reliable travel for everyone, whether they walk, bike, take public transit, or drive

- Build on the current transit services through new and enhanced commuter rail, light-rail, and bus services
- Bring together better transit and travel options for people to explore communities without relying on a car
- Include micromobility strategies, rideshare, and microtransit options that would make first- and last-mile options safer and more convenient

In the following selection of case studies, you will learn how an international airport uses GIS to plan for future passenger demand while supporting and encouraging local and regional economic growth. You will see how a state department of transportation uses GIS to make best-value investment decisions by evaluating the regional terrain and weather to meet and respond to safety and operational challenges. In another story, a state department of transportation shifts strategy, placing extra emphasis on climate resiliency within its long-term planning and operational processes. Finally, one of the largest transit agencies in the US expands the use of GIS across all facets of its business and services.

PREPARING FOR THE FUTURE

Reno-Tahoe International Airport

AIRPORTS PLAY AN INCREASINGLY IMPORTANT ROLE IN connecting economies and global supply chains and in driving economic development in their regions. More than 4 million passengers a year travel through Reno-Tahoe International Airport (RNO), and the airport delivers more than $3 billion in economic impact for the region. RNO is owned and operated by the Reno-Tahoe Airport Authority (RTAA).

With passenger travel forecast to exceed more than 6.5 million passengers per year, RNO must consider not just accommodating future passenger growth, but also how the airport could expand its positive impacts on the Reno-Tahoe region. Capitalizing on the fact that an efficient and well-connected airport can deliver a competitive advantage to the region, the airport is a central distribution hub to the western states to help drive greater economic development in the region.

To receive Capital Improvement (CIP) funds from the Federal Aviation Administration (FAA), RNO must also comply with the FAA's Airport GIS program. For example, RNO needed to collect the digital data required to support the transition to a satellite-based air traffic control system called NextGen, which required collecting precise survey information of airfield data in GIS format. RTAA set aside $1.25 million to help fund the project and ensure that the airport moved forward in a smart but comprehensive way.

To collect the required digital information, the airport partnered with Arora Engineers Inc. and began collecting precise survey and ortho-imagery data using ArcGIS Online to deliver both data products and applications to the airport. The data also supported wider airport and airspace operations, including unmanned aircraft systems restricted airspace, airport height restrictions, airport land use, emergency flood analysis, and airport surface radar coverage analysis.

While survey crews were collecting data, RTAA and the Arora team were meeting with the property division, facilities and maintenance, engineering and planning, air service, and operations and finance departments to help define strategic business applications, including a GIS portal through which all the major applications and data could be accessed. The GIS portal also serves as a basic data viewer, allowing airport staff to view relevant locations on the airfield and within the terminal. Staff use the data viewer to generate their own high-quality maps for meetings and reports.

Another GIS application includes a map-based document discovery system, in which text and spreadsheet documents, photos, CAD drawings, and maps are geographically tagged to a location and retrieved through a map interface. Similarly, another GIS application is used to capture and organize all the terminal lease spaces, which integrates with the airport's property and leasehold management application, allowing staff in the property division to access any of the lease information and understand the status of every property within the terminal, airside, or off airport facilities.

In addition to using GIS for business planning and meeting FAA requirements, RTAA also uses GIS to develop long-range planning for the airport and coordinating with subcontractors. The master plan for the airport focuses on expanding commercial air service, general aviation and cargo service, and regional economic development. With an emphasis on customer service, RTAA is strengthening a true partnership between the airport and the community that encourages economic growth.

Reno-Tahoe airport's spatial data portal.

The Reno-Tahoe airport's spatial data portal is a web application that provides access to maps, additional GIS applications, and data.

GIS data is central to the development of the airport's master plan, which calls for a $1.6 billion modernization and capital improvement plan to support growing passenger and cargo volumes. ArcGIS Online is used to support public participation and gives RTAA the ability to add subcontractors to the master plan team. GIS maps and applications provide all stakeholders with timely and consistent information throughout the planning review process.

This story originally appeared as "Discover How GIS Fueled Their Efforts towards Success: A New Approach at Reno-Tahoe International Airport" on esri.com, 2018. All images courtesy of the Reno-Tahoe Airport Authority.

INVESTING IN THE JOURNEY

Colorado Department of Transportation

THE COLORADO DEPARTMENT OF TRANSPORTATION (CDOT) is charged with maintaining road infrastructure and—most importantly—with improving safety and the journey experience of travelers. CDOT's vision is to enhance the quality of life and the environment by creating an integrated transportation network.

CDOT uses ArcGIS software to collect information from many sources and providers into a single integrated information source that can be accessed and viewed simply and quickly via common browsers and applications. Staff can use, add to, and guide information development from within their organization and with partners and providers. The consequent removal of institutional barriers results in performance improvements across the board.

"What that means for CDOT is that business decisions regarding when and where to invest in infrastructure maintenance and improvement can now be led by the head and not the gut," explained chief engineer Joshua Laipply. "Historically, many of our field personnel's investment recommendations were based on anecdotal evidence—their 'feel' for what was going on. Now, decisions are much more intelligence led. As an example, because we can overlay detailed information on specific weather events and actions that were taken, we know precisely where snowplows have salted the roads. We can gain a far better idea of the likely locations of pavement deterioration caused by chemical processes and where we might need to intervene. We're shifting to a more proactive stance in infrastructure management and advancing operational excellence."

CDOT's data enrichment effort began with an evaluation process that included a detailed review of existing data and data quality. The GIS dashboards, map viewers, and mobile apps developed by

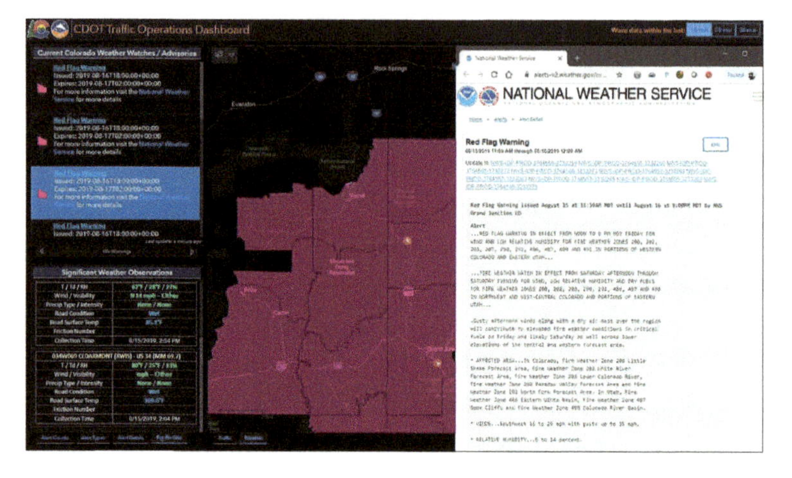

The Colorado Department of Transportation traffic operations dashboard provides critical weather watches and advisories.

CDOT were used to identify and address data gaps and redundancy and improve data quality.

In addition to improving processes from an asset management perspective, CDOT also uses data from sensors and smart devices to improve real-time and predictive traffic and incident management performance, including information from first responders, tow trucks, and the automatic vehicle location systems on winter maintenance vehicles.

"Road networks are becoming information networks," Laipply said. "But more data doesn't necessarily make you smarter. It's about how you pull it together and visualize things. At the same time, it's important to acknowledge that opportunities for improvement could be missed if excess data is discarded."

Chief data officer Barbara Cohn sees the ability to readily prove and support good business use cases as critical to CDOT's continuing development and success. Cohn notes the ability to draw on information from disparate sources, including those from outside the organization. To complement its traffic management operations, CDOT worked with Esri to access data streams from private-sector providers, such as Waze and HERE, and look further into the future by overlaying mapping information for new property developments, freight movements, and demographics to predict and effectively manage traffic patterns.

"The ability to visualize vast amounts of information means that our understanding of what's happening is much more acute. Visualizations help identify trends, distill increasing amounts of information, and transform raw data into actionable insight," Cohn said. "Having the opportunity to look at all data, not just some, is inspiring new, creative, and cooperative problem solving."

Colorado's geography presents a variety of safety and operations-related challenges. The Rocky Mountains are in the western half of the state, and CDOT must manage more than 35 mountain pass roads, which are impacted by heavy snow and susceptible to avalanches during winter.

"2018 was a record year for avalanches, and we needed a way to track which avalanche paths were most likely to cause disruption. We also needed to know where to plant and, if necessary, recover explosives," Laipply explained. "By using a GIS mobile app, our field operatives no longer have to carry with them 10 years of handwritten notes. Via a smart device, they have all that information at their fingertips, and our reactions to events are far faster and more effective."

"That's what's getting people very excited," Cohn said. "The GIS technology is reusable, and the apps are configurable, requiring little or no custom code. Because we can fail fast and fail small, people at all levels are becoming more creative. By democratizing data, we're getting to that 'boots on the ground' knowledge that is so important."

The Colorado Department of Transportation night inspection dashboard is used to monitor roadway asset defects.

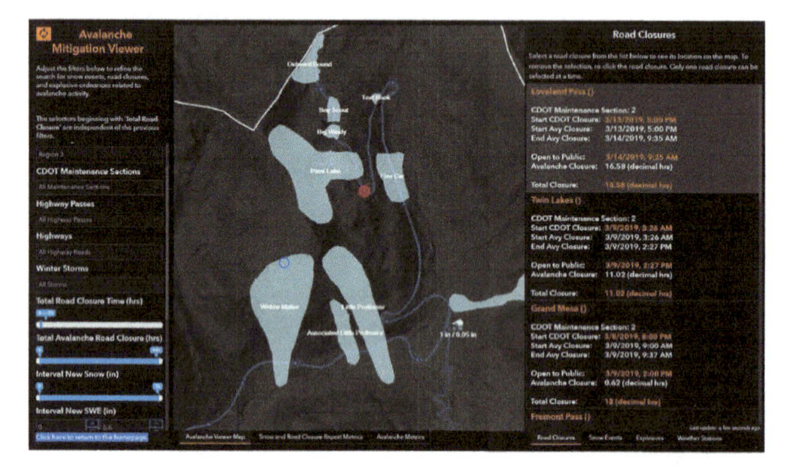

The Colorado Department of Transportation avalanche mitigation viewer is used to track snow events, road closures, and the use of explosive ordnance for avalanche prevention.

In addition to being able to make best-value investment decisions, CDOT is seeing an uplift in its real-time traffic management operations, including tactical improvements to performance and better insights about potential long-term gains, such as better road design.

This story originally appeared as "Colorado Department of Transportation - Esri System of Engagement" on esri.com, 2018. All images courtesy of Colorado Department of Transportation.

BUILDING INFRASTRUCTURE RESILIENCE

Vermont Agency of Transportation

CATACLYSMIC EVENTS HAVE AN UNERRING ABILITY TO FIND weaknesses and shortcomings in any human-made system and precipitate great change to those systems. To prepare for climate-related weather events, the Vermont Agency of Transportation (VTrans) developed a Transportation Resilience Planning Tool (TRPT). At its heart, the tool uses dashboards and apps based on ArcGIS technology.

In Vermont's case, precipitation is no idle choice of word. The state's pivotal event was 2011's Hurricane Irene, which deposited as much as 11 inches of rain in parts of the state within hours, causing widespread flooding.

Irene destroyed or damaged more than 2,400 roads, 800 homes and businesses, 300 bridges (including historic covered bridges), and half a dozen railroad lines in Vermont. Several communities were entirely cut off by the flood and erosion damages, some for up to two weeks, and disruption to US Route 107 and other roads across the Green Mountains made east–west travel through the southern half of the state nearly impossible.

Though Irene struck in August 2011, repairs went on for years. In total, damages in Vermont from America's sixth-costliest hurricane came to $733 million.

Goals in developing the TRPT included the ability to identify at-risk infrastructure, consider building greater resiliency into infrastructure whether through initial design or subsequent upgrades, and apply more effective management strategies based on key safety and capacity access points along the transportation network.

"We have a declared disaster in the state around every 18 months or so. That's not necessarily always on the scale of Irene, but we want to get ahead of the problem next time by identifying vulnerable network segments and how critical they are. That gives us a measure of risk and enables us to prioritize changes as well as recovery efforts," said Joe Segale, policy, planning, and research director at VTrans.

Before Irene struck, Segale says resilience was not as high of a priority. "There were programs for safety, for bicycle and pedestrian, bridges, transit ... but not resilience. Irene changed that," he said.

Now, the agency places more emphasis on strategic and long-term planning with climate resiliency built in. The TRPT will provide the data the agency needs to increase resilience as part of planning and prioritization.

"The way you bring resilience forward is to integrate it as part of what you do every day, not have it as a stand-alone program," Segale said.

From the outset, GIS was regarded as the solution of choice. Segale points to the ability to bring together the spatial elements of VTrans road infrastructure and local river systems.

"There's just no other way to do it, in my mind—you can't just use a spreadsheet," he explained. "Another reason is ease of use once the data gathering and app building are done. Vermont has around 250 small towns, and while we have hundreds of engineers within the department who can interpret data, we also have a lot of volunteers throughout the state who need a tool that allows them to easily understand where the risks are."

VTrans has spent around $1 million on the TRPT so far, with the initial half million dollars 60 percent funded by the Federal Emergency Management Agency (FEMA). It is a clean-slate exercise—nothing remotely similar was in place before—and is close to completion, except for final data development in the remaining parts of the state.

A pilot on three watersheds included application of the TRPT on the Route 107 and Route 9 east–west corridors. Upon completion of data gathering for the whole state, the tool will form a major part of Vermont's flood resilience efforts. Segale says it will significantly help VTrans partners and regional planning commissions develop local hazard mitigation plans and access FEMA funding. Since Irene, VTrans has been collaborating with a key partner, the Vermont Agency of Natural Resources, Department of Environmental Conservation, Rivers Program. Their joint ambition is to reduce flood and erosion risks while also protecting water quality and habitat.

Roy Schiff of SLR consulting firm says that determining vulnerability—how susceptible an asset such as a bridge, road, or culvert is to failure—once relied on hindsight rather than a forecasting process. A statewide geomorphic dataset, which covered roughly 2,500 miles of field data on river forms and functions, was merged with historical data on damage to roads and infrastructure.

"We looked at the proximity of the two," Schiff said. "So how much of a road is in a floodplain or river corridor? What's the vertical difference? How does the width of a bridge compare to the width of a river? We developed screening methods to identify vulnerabilities, and then developed algorithms for inundation, erosion, and deposition, and for three floods—10-, 50-, and 100-year."

Whereas other states' solutions tend to concentrate only on inundation and FEMA's flood maps, Vermont's stands out because it also considers the other two processes—erosion and deposition.

In terms of criticality modeling, what is truly novel is how the network criticality indices have been developed and refined. In simulations, vulnerability probabilities are attached to road network segments to determine failures.

"We found that the modeling could identify spots that, if they broke, the whole road network started functioning poorly," Schiff

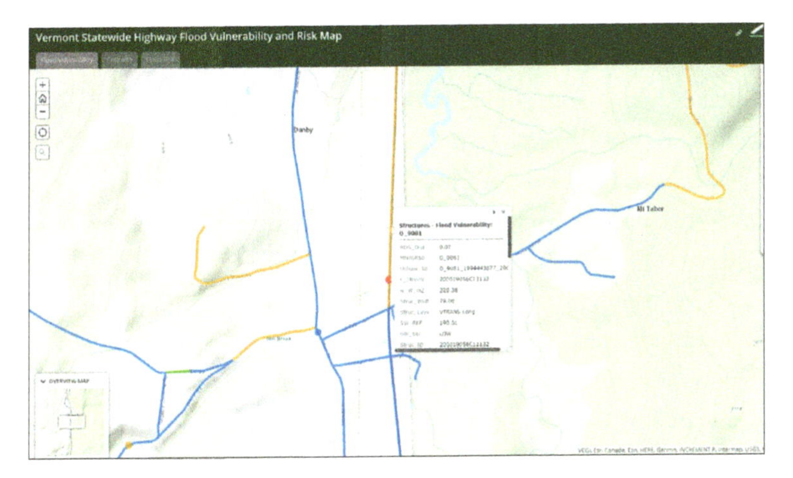

The Vermont Statewide Highway Flood Vulnerability and Risk Map application evaluates every transportation asset based on flood vulnerability, criticality, and flood risk.

said. "But that left out some areas that people really relied on to get around when parts of the network were down. So we asked people to identify the locally important roads."

Schiff and his team also employed a Critical Closeness Accessibility Index developed by Jim Sullivan at the University of Vermont.

"This looks at how close people are to important services—shelter, fire, rescue, food, schools, and so on," Schiff said. "It asks, 'If this part of the network breaks, are we going to lose connection to services?' Bringing all these elements together gives a criticality value. The risk value is an average."

At the earliest development stages, members of the public, along with staff from VTrans and the planning commission, were included in the user group and helped bring out specific knowledge.

"We asked for input on whether we got the vulnerabilities right or if there were things that we missed," Segale said. "On the criticality side, we asked if there were roads we were not picking up with our methodology. We also needed the app to be easily understandable by them—and especially by those who make budgetary decisions—so that it feeds into their capital planning."

Barbara Patterson, of Stone Environmental, was responsible for TRPT application development and noted usability as a prime criterion.

"On opening, the app is already displaying the vulnerabilities of road segments," Patterson said. "This happens by default at the highest level. Finding 'your' watershed is relatively simple, and you can search by location."

The app is now operationally complete, and the team is working to build out the dataset statewide. Regional planning commissions have been supportive of the tool and started doing their own analyses. "In particular, they're really excited to see all the work that they've done accessible in one platform," Patterson said.

"The tools developed by VTrans give a measure of where we are in terms of flood resilience," Segale said. "We can determine what percentage of our culverts are highly vulnerable. A particular lesson from Irene is that culverts were the 'forgotten soldiers,' but there are all kinds of performance metrics that we can pull out and track over time."

Though Segale acknowledges the difficulty in finding the right data to measure what's most important, he is confident that the new tools will provide that information.

Next steps could include further expanding the stakeholder group to create yet more synergies, a more comprehensive dataset, and a better appreciation of performance for roads and other infrastructure.

For example, Green Mountain Power, which supplies around three-quarters of Vermont's electricity, has expressed interest in the VTrans tool and methodology as a means of gauging power

Vermont Agency of Transportation uses GIS to evaluate and report the vulnerability, criticality rating, and proposed remediation measures for assets such as bridges, shown with the cyan dot.

infrastructure vulnerability. The state's railroad infrastructure could also benefit similarly.

Another possible application is to overlay results from the TRPT with water quality improvement needs on the highway system to look for opportunities to address both issues. Vermont's Lake Champlain, one of the largest by volume in the US, is currently the subject of a pollution prevention, control, and restoration plan. Agricultural and urban runoff has led to a phosphorus problem and an issue with algal blooming. One of the river basins currently monitored by VTrans, that of the Missisquoi, feeds directly into the lake. Segale says it may be possible to overlay the resilience tool to determine any relationship to water quality. Culvert performance at wildlife crossings could also provide an opportunity for addressing multiple issues.

Finally, the TRPT provides a strong basis for MAP-21 Part 667 reporting to the US Federal Highway Administration (FHWA) of repeat damage to the National Highway System—a mandated requirement. Its benefit is threefold: providing a greater understanding of where repeat damage is a result of declared emergencies, ensuring that assets are constructed so that they do not suffer repeated damage, and safeguarding FHWA funding so that it is used where it has the greatest benefit.

Vermont has a proud tradition of responsible environmental stewardship. With the creation of the TRPT, VTrans takes a leadership role, building climate resiliency into its statewide planning processes and helping local governments prepare for a much less predictable future.

This story originally appeared as "Preparing for Weather Disasters: Vermont Builds Resilience into Infrastructure Plans" on esri.com, 2018. All images courtesy of the Vermont Agency of Transportation.

BUILDING SMARTER, MORE RESILIENT

Colorado Department of Transportation

THE COLORADO DEPARTMENT OF TRANSPORTATION (CDOT) uses GIS tools to build a strong foundation for its resiliency efforts, preserving and rebuilding infrastructure in the most strategic and cost-effective manner.

After severe-weather events, the initial inclination may be to repair or rebuild damaged infrastructure, but if the original design was lacking, the repair ensures potential failure. Additionally, there are costs beyond those of reconstruction to consider. Some road sections and bridges are so critical to a geographic area's economic and social well-being that their speedy restoration to full operational capacity is a necessity. Other roads and bridges are of secondary or tertiary importance, and while being out of service may be inconvenient, their reconstruction can be delayed if necessary.

Determining the true value rather than just the cost of restoration work is something CDOT has been doing with the support of ArcGIS® tools from Esri. The aim is to bring greater coherence to resiliency planning and make resilience part of the day-to-day activities of CDOT.

CDOT's resiliency program manager, Lizzie Kemp, says a catalyst for the department's resiliency planning came when staff realized that many of the state's road facilities that were destroyed by flooding in 2013 had also been wiped out by a flood in 1976. The department was also motivated by a federal requirement for the periodic evaluation of facilities repeatedly damaged during emergency events.

In the aftermath of the 2013 storms, then Colorado governor John Hickenlooper set up the multisector and statewide Colorado Resiliency Office (CRO). Rather than rely on the Federal Highway Administration's (FHWA) Emergency Relief Program for Federally

Owned Roads and wait for things to go wrong before taking remedial action, the CRO was meant to be proactive in terms of resilience.

In keeping, CDOT took a strongly proactive approach to resiliency, establishing its own dedicated Resiliency Working Group (RWG) and Executive Oversight Committee (EOC). Established in 2015, a tenet of the RWG is that its influence should be from the bottom up and not from the top down. CDOT's EOC meets every other month and the RWG meets monthly. From the beginning, both have enjoyed strong, regular participation from stakeholders at all levels. After several years of successful operation, their work was formalized in a 2018 state-level policy directive.

CDOT's first proactive look at risk and resiliency focused on assets and potential hazards across Colorado on Interstate 70. This route forms Colorado's main east–west corridor between Utah and Kansas, and parts of the interstate's Colorado section are among the most challenging in engineering terms. The dual-bore, four-lane Eisenhower Tunnel, for instance, passes under the Continental Divide in the Rocky Mountains and is the highest point on the US Interstate Highway System. Then there is Colorado's climate to consider. This might best be described as complex and its effects on road infrastructure as profound—2013's storms saw more than 400 miles of roads rendered inoperable.

The I-70 project examined both the risk of climate-related damage from CDOT's perspective, but also in terms of the consequences to road users—the traveling public and the impact on freight operators. Analysis used a variant of the seven-stage Risk Analysis and Management for Critical Asset Protection (RAMCAP) methodology to examine a wide range of threats, including floods, avalanches, and rockfalls.

The goal of the project was to develop a cost-benefit analysis to guide CDOT staff in their maintenance and infrastructure planning.

The methodology, following the urging of the Federal Highway Administration, was focused on determining the quantitative costs and benefits of various investments designed to harden Colorado's highway infrastructure. The analysis set out to determine the most critical highway assets on the I-70 corridor and develop a criticality model maximizing the triple (social, economic, and environmental) bottom-line considerations.

To determine criticality, the study looked at not only total traffic volumes but also the value of freight carried, as well as the impacts on tourism. A total of 35 socioeconomic variables were used to determine how quickly communities would rebound from an event as well as what the influence (if any) of alternative routes would be. GIS was the underpinning technology used, providing a dependable predictive tool.

The study helped establish that more than 20 percent of the Colorado portion of I-70 should be rated highly critical. In the cost-benefit analysis, CDOT sought to answer two essential questions: "How much would it cost to buy down the risk?" and "Is it worth the investment?" From the I-70 analysis, CDOT calculated the cost of its own annualized risk at $6 million, most of it related to rockfall, with flooding as an important secondary risk. But the cost of the annualized risk to road users was significantly higher—more than $164 million—with 75 percent of the cost related to road closures from flooding, and with rockfall risk coming in second.

The initial study found that a significant portion of risk and potential damage could be attributed to nonperforming culverts. As Kemp notes, culverts are a lesser-cost item, but they represent an opportunity to achieve significant resiliency improvements at relatively little cost.

Five or six years before the 2013 flooding, CDOT did not have a thorough understanding of where its 65,000 culverts were, much

Colorado Department of Transportation uses GIS dashboards to track culvert locations and conditions along state highways and roads.

less an appreciation of their condition. It was, says Kemp, a "massive effort" to get out there, document them all with GIS apps from Esri, and rate each as good, fair, or poor. Now, there is a far greater understanding of each culvert's location, condition, and importance. As a result of the I-70 project and subsequent work, decision-making in terms of expenditure on culvert maintenance and repair is data driven and supported by one of several ArcGIS Dashboards.

The I-70 project involved a data-intensive, GIS-based effort, with many of the calculations conducted by an outside consultant. Kemp and other CDOT staff sought to bring the same methodology and expertise in-house so CDOT could replicate the methodology statewide. Using internal resources, Kemp wanted to create what she calls "a cookbook for how to calculate risk."

That cookbook was eventually published in August 2020. It includes methodologies for calculating a threat paired with each asset (flood with roadway, flood with bridge, etc.). These methodologies have been moved into a spreadsheet environment, which greatly

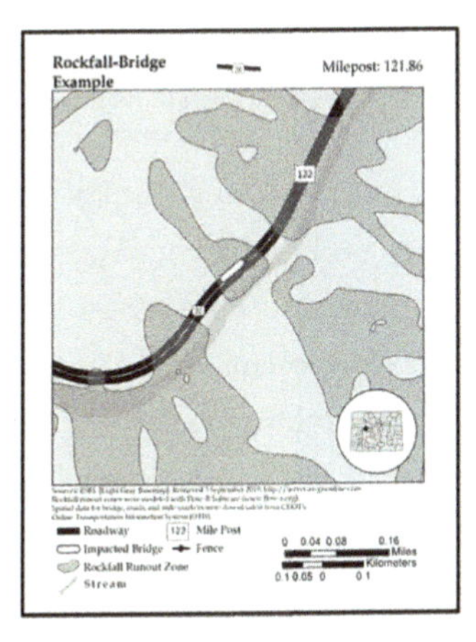

Colorado Department of Transportation combines GIS analysis, travel demand modeling, and local knowledge from traffic engineers to evaluate the full impact of potential road closures.

simplifies the data entry and calculation processes, with a future goal to build batch-processing capabilities to analyze many assets at once.

Water volumes at specific locations were calculated with the US Geological Survey (USGS) StreamStats application, and Kemp worked with CDOT's travel demand modeling specialists to determine the effects of alternative routing for any roadway that would be blocked by a weather-related disaster. Local knowledge from traffic engineers helped ensure that information was complete in terms of validating calculated detours. Closure effects were assessed on a link-by-link basis and in both directions to understand the full impact of a potential road closure.

To analyze each asset and combined threat, the model was run twice. The first run established baseline condition and cost, while the second run analyzed the reduction in risk achieved by proposed mitigation. Taken together, the two model runs established the benefit-cost ratio. Specific outcomes included predictions, for instance, that a rockfall in the Glenwood Canyon section of I-70 would result in travelers making a 187-mile, four-hour detour. Beyond some of the straightforward calculations, however, are more sobering statistics.

According to the Federal Emergency Management Agency, blocked and closed roads have a direct impact on businesses' survival—40 percent of local businesses never reopen after a disaster, and 90 percent of small businesses fail within a year if unable to reopen within five days. The costs of delays to shippers can be equally devastating: depending on the product, a delay can cost shippers between $25 and $200 per hour of the event duration. The result of Kemp's calculations is that, on average, every $4 spent on proactive resiliency saves at least $25 in repairs, and the message is clear: preparation makes good sense.

This was just the start for Kemp, however. She has a working list of projects to take resiliency forward within CDOT. High on her list is to help the agency develop resiliency performance measures, and to update CDOT's resiliency manuals. The criticality analysis has led the agency to investigate new redundancy measures, including a statewide detour map and better ways to gauge the impact to remote communities.

Finally, Kemp is in the process of completing five case studies that she hopes—in continuing the cookbook analogy—will help bake resiliency into various aspects of CDOT's day-to-day business processes. Kemp indicates that the case studies should be completed and available by the end of 2021.

The result will be a matrix of mitigations that, according to Kemp, highlights how organizations need to be just as resilient, with backup servers and cross-trained staff, as the assets they manage. Kemp says it comes down to being able to make robust decisions, confident in the knowledge that they can be justified. If, for instance, an asset is noncritical, is there any need to harden it? And if a more critical asset is damaged in a weather-related disaster, can staff demonstrate that rebuilding it better makes good long-term financial sense?

This story originally appeared as "Colorado DOT: Building Back Better and Smarter for a More Resilient Transportation System" on esri.com (2019). All images courtesy of Colorado Department of Transportation.

TAKEAWAYS

GIS IS AN ESSENTIAL TOOL FOR TRANSPORTATION planning. Data-driven decision-making is the steel thread that binds transportation plans to communities and the business of moving people, goods, and services. GIS brings complex data into focus, making the story clearer, providing actionable intelligence to leadership, and delivering better communication between agencies and the customers they serve.

Key takeaways for applying GIS to transportation planning

- **Effective collaboration:** GIS maps and applications enable consensus building within the organization and help communicate planning concepts, which strengthen decision confidence and earn greater public support.
- **Business growth:** GIS analysis is critical to understanding growth potential and designing effective programs to increase traffic and grow markets. It can also track customer lifestyle and demographic characteristics and mobility patterns.
- **Regulatory compliance:** With GIS, planners can visualize and analyze environmental impact on communities, natural environment, and affected populations and generate validated plans to help prepare compliance data for regulatory agencies.
- **Strategic property management:** GIS mapping and analysis provides a way to visualize and understand property locations, optimize space and facility use, analyze energy consumption or lease potential, and mitigate risks to property and facilities.

- **Environmentally positive action:** With GIS, transportation agencies and companies can combine and analyze authoritative data for biological and environmental studies and assessments and compare alternatives to make the best decisions.
- **Operational resiliency:** GIS analysis can help predict possible futures, allowing transportation agencies to understand and plan for climate change impacts on infrastructure—such as sea-level rise or increases in precipitation, storm intensity, or temperature extremes—and be better prepared for emergencies and incident response.

Learn more about applying GIS to transportation planning and sustainability in the last section of this book, "Next Steps."

NEXT STEPS

GIS IS AN ESSENTIAL TOOL FOR TRANSPORTATION professionals. With GIS, transportation agencies and companies can build a strong sense of location intelligence throughout the organization, improve communication, and make working together easier and mutually beneficial. The case studies presented in this book show how transportation agencies are using GIS today to become more efficient, safer, and sustainable. As a next step, the reader may want to explore different ways to apply GIS to their organization and, ultimately, experience GIS technology firsthand.

The following information is organized by the topics covered in this book and represents some of the most common strategies that transportation agencies use to apply GIS to their organization.

Operational efficiency

The objectives of improving operational efficiency are making tasks more efficient while minimizing risks to the transportation system, ensuring that the system is flowing at optimum levels, and keeping costs down or avoiding unnecessary costs. With GIS, transportation personnel can use location intelligence to fine-tune daily workflows and coordinate tasks. GIS analysis provides fresh insights about performance, risks, resources, and costs that are not readily apparent in asset management data. Using location, transportation agencies can discover patterns and trends that simple reporting cannot detect, thereby improving operational efficiency and transparency by increasing location intelligence throughout the organization. Use the following checklist to increase operational efficiency.

Operational efficiency checklist

- ☐ Transition away from cumbersome legacy processes and inefficient paper-based workflows by using GIS maps and mobile apps to create asset inventories and perform inspections.
- ☐ Improve data collection abilities and workforce mobility using GIS apps on smart devices, drones, and sensors to increase data timeliness and accuracy while doing maintenance projects and responding to citizen complaints.
- ☐ Accomplish faster data analysis and mapping that support decision-making and mobile workers in daily operations.
- ☐ Increase response effectiveness for citizen requests and emergencies using real-time GIS dashboards to monitor and share performance metrics and visualize trends.
- ☐ Create better opportunities for productive participation within the community by sharing GIS maps, dashboards, and reports with civic leaders, business owners, and the public.

Safety and security

Preparing for safety and security issues involves two major strategies: identifying infrastructure vulnerabilities and improving collaboration with other departments and partner agencies, such as local police and fire departments. GIS mapping and analytics allow transportation organizations to detect where potential hazards threaten the integrity of infrastructure in different scenarios. GIS maps and applications are also effective communication tools for managing, sharing, and using data during an incident. Use the following checklists to strengthen transportation safety and security preparedness, response, and recovery efforts.

Safety and security preparedness checklist

- ☐ Craft organization-specific operational workflows and use-cases based on different scenarios relevant to areas of operation, and identify the decisions that are needed and what datasets, including location datasets, support these key decisions.
- ☐ Review interagency workflows based on the scenario; identify the key decisions, contacts, roles, and responsibilities for the datasets; identify or create the data-sharing agreements between partners; and start collecting and sharing the data identified for different scenarios.
- ☐ Create hazard maps showing the locations of safety risks and security vulnerabilities—such as fueling stations, entrances, and exits, and security checkpoints—that could potentially impact workers, customers, and vendors.
- ☐ Map and identify vulnerable populations based on risk factors to help the organization monitor at-risk groups and geographic regions adjacent to operational properties and infrastructure.

Safety and security response checklist

- ☐ Create real-time GIS dashboards to support decision-makers and responders during incident response. Dashboards are also an efficient and effective way to keep partner agencies, the public, media, and local and national leadership apprised of current and changing conditions.

- ☐ Deploy GIS damage assessment apps to response and repair crews as quickly as possible to determine where and when to deploy additional response and recovery resources.
- ☐ Share public information maps with the public and media to report the status of active incidents, show areas affected by incidents, and alert people to changing conditions.

Safety and security recovery checklist

- ☐ Create recovery maps to show where incident recovery efforts are ongoing and planned and where staff and nearby populations in need can find resources.
- ☐ Build a safety and security destination, such as a web-based recovery hub, which serves as a single source of information for all recovery efforts. A GIS hub can be a place to share maps, resources, and recovery initiatives.
- ☐ Collect and organize data, including maps and spatial analysis, to create a comprehensive review of recovery needs that can be used to request financial assistance and other funding support.

Asset management

Enabling asset management software and databases with location intelligence helps transportation agencies better understand where assets are, how those assets are related to other assets, and spatial patterns and trends about the conditions and performance of those assets overtime. When GIS and asset management systems are combined, the distributed and interconnected nature of asset locations becomes apparent, helping the organization process, synchronize, and share information in an efficient and effective manner across departments. Use the following checklist to reinforce transportation asset management.

Asset management checklist

- ☐ Produce a more accurate asset inventory using GIS data collection applications, and combine location intelligence into existing asset management systems and inventory tracking.
- ☐ Make asset documents, including engineering diagrams and schematics, available to inspectors and repair crews and allow automatic updates to the asset management system via GIS mobile applications.
- ☐ Create GIS dashboards to monitor asset conditions in real time and analytic maps to visualize performance patterns and trends by location and time, forecast maintenance costs, and improve reliability.
- ☐ Analyze property locations to evaluate land use, maximize revenue potential, optimize space and facility use, and track energy consumption by facility and area.

Planning and sustainability

Transportation agencies use GIS to analyze and evaluate projects, taking into consideration the long-term sustainability and resiliency of the infrastructure, while projecting anticipated growth in business or demand and the changing needs of customers. GIS mapping and analysis provides a geographic perspective that improves understanding of current conditions and existing stresses on transportation systems, including the future needs of customers and businesses, economic development patterns, and state and federal requirements. Use the following checklist to enhance transportation planning and sustainability.

Planning and sustainability checklist

- ☐ Create, track, and review development projects with a digital twin of the transportation system by combining GIS with engineering design data, building information modeling, imagery, and 3D information to give planners and developers a common view of transportation projects and plans.
- ☐ Visualize transportation system project scenarios based on population change, economic growth, projected demand, employment densities, land use, and compliance.
- ☐ Use mapping and spatial analysis to balance transportation plans and projects with economic, social, and environmental considerations to maximize the value to communities and customers and achieve sustainable growth.
- ☐ Guide route optimization with location-based accessibility tools and travel pattern data to maximize ridership and freight and cargo capacities.
- ☐ Improve communication during the transportation planning process using GIS maps, dashboards, and a hub website to build consensus across organizations and grow public confidence by sharing strategic and cost-effective decisions.
- ☐ Monitor and manage the transportation construction projects using GIS dashboards and mobile apps.

HOW TO GET STARTED WITH GIS

THERE ARE TWO WAYS THAT YOU CAN START APPLYING GIS to transportation operational efficiency: Esri solutions and hands-on learning.

Esri solutions for transportation are designed for quickly applying GIS maps and apps for these purposes:

- **Traffic crash analysis:** The Traffic Crash Analysis solution provides a set of capabilities that help you analyze crash data using methodologies defined by the US Road Assessment Program and the Federal Highway Administration, including the ability to identify where concentrations of serious injury and fatal crashes occur and then share the results with decision-makers and the public.

- **Transit outreach:** The Transit Outreach solution allows you to identify and understand community needs and incorporate these needs into transportation plans and programs. Transit Outreach is typically implemented by transit agencies that want to become more transparent and responsive and ultimately provide better transit service.

Hands-on learning will strengthen your understanding of GIS and how to use the technology to improve operation efficiency. A good place to start is Learn ArcGIS. Learn ArcGIS is an online collection of free story-driven lessons that allow you to experience GIS applied to real-life problems. With Learn ArcGIS lessons, you will gain familiarity with the following:

- **Planning and sustainability:** Plan the expansion of a public transit network to help connect predicted housing growth in a downtown area.

- **Asset management:** Create a public transit network dataset for buses and subways, locate new facilities, and determine important destinations for riders, such as jobs, education, shopping, health care, and recreation.
- **Safety and security:** Use 3D visualization and spatial analysis to identify potential aviation obstructions and danger to aircraft operating in the area.
- **Planning and sustainability:** Find businesses in specific industries located at various distances from railway lines, present the information to executives, and launch a campaign targeted to nearby businesses.
- **Safety and security:** Make a severe-weather map to track potential hazards to a transportation system.
- **Safety and security:** Analyze fatal cycling accident data to determine which areas and routes are safest or more dangerous for cyclists.
- **Safety and security:** Collect location data to increase preparedness for natural disasters.
- **Safety and security:** Create situational awareness for all the emergency events in an area and track fire, police, and medical responses in real time.
- **Operational efficiency:** Use GIS mobile applications and dashboards to manage multiple stages of asset replacement, including data collection, inspections, and replacement records.

Learn about additional GIS resources for transportation by visiting the web page for this book:

go.esri.com/mf-resources

CONTRIBUTORS

April Blackburn
David LaShell
Jeffery Peters
Monica Pratt
Barbara Shields
Steve Snow
Citabria Stevens
Carla Wheeler

ABOUT ESRI PRESS

AT ESRI PRESS, OUR MISSION IS TO INFORM, INSPIRE, AND teach professionals, students, educators, and the public about GIS by developing print and digital publications. Our goal is to increase the adoption of ArcGIS and to support the vision and brand of Esri. We strive to be the leader in publishing great GIS books and we are dedicated to improving the work and lives of our global community of users, authors, and colleagues.

Acquisitions

Stacy Krieg
Claudia Naber
Alycia Tornetta
Craig Carpenter
Jenefer Shute

Editorial

Carolyn Schatz
Mark Henry
David Oberman

Production

Monica McGregor
Victoria Roberts

Marketing

Mike Livingston
Sasha Gallardo
Beth Bauler

Contributors

Christian Harder
Matt Artz
Keith Mann

Business

Catherine Ortiz
Jon Carter
Jason Childs

For information on Esri Press books and resources, visit our website at esri.com/en-us/esri-press.